The Quality Brew

Coffee is a major commodity of export in developing countries. In India, it generates significant revenues besides providing employment to many. This book critically looks into various aspects of the coffee production and marketing industry and its potential for promoting sustainable development, poverty alleviation and providing decent work. The small-holder-dominated coffee sector is under unprecedented crisis attributed to a host of factors like declining productivity, incidences of new pests and diseases, adverse price trends and issues related to global warming-induced climate change. Supported with primary survey and secondary data, this book explores challenges related to coffee production and its economic importance in India. It analyzes factors affecting the productivity of Robusta and Arabica varieties of coffee and offers important inferences about the relationship between age and productivity of the plants and factors that guarantee high productivity. The book also sheds light on institutional interventions and the use of IT applications, along with online trading, in the coffee sector. Comprehensive and insightful, the book will be of interest to students and researchers of agriculture studies, agricultural economics, horticulture with a special focus on spices and plantation, sustainability studies, development studies and crop science. It will also be useful to stakeholders of plantation crops, such as growers as well as policymakers, processors, traders and exporters at different levels.

Gana Shruthy M.K. is currently working as Assistant Professor, Department of Collegiate Education, Government of Karnataka, and serving as Special Officer at Karnataka State Higher Education Council. She had previously worked at the Institute for Social and Economic Change (ISEC), Bangalore. The UGC has conferred her with Junior/Senior Research Fellowship Award. She was also conferred with Senior Research Fellow (SRF) in Agricultural Economics by NAARM, Hyderabad. She is also a rank holder in the state-level entrance exam for the post of assistant professor conducted by the Karnataka Examination Authority (KEA) and was selected for the post of assistant professor in economics. She has published in reputed national and international journals. She obtained her doctoral degree from the Institute for Social and Economic Change (ISEC), Bangalore. Her area of specialization includes agricultural economics.

Parmod Kumar is presently Director at the Giri Institute of Development Studies, Lucknow, while he is on deputation from his position as Prof & Head, Agricultural Development and Rural Transformation Centre, Institute for Social and Economic Change, Bengaluru, India. He was Director, Giri Institute for Development Studies, Lucknow during 2021 to 2024. He has previously worked at the National Council of Applied Economic Research, New Delhi, and the Institute of Economic Growth, Delhi. He obtained his post-doctorate as Sir Ratan Tata Fellow and doctorate from JNU, New Delhi. He was a fellow under the *International Visitors Leadership Program* sponsored by the US government. He has been Visiting Scholar at Glasgow University, OECD-FAO Paris, Sichuan Academy of Social Sciences, Chengdu, and Zurich University of Applied Sciences (ZHAW), Switzerland. He has authored/co-authored more than 14 research volumes and published more than 80 research articles in refereed national and international journals. He was the managing editor of the *Journal of Social and Economic Development* (JSED), Springer, and, at present, he is Member Editorial Board of the *Indian Journal of Social Justice*. He is also an executive member of the *Indian Journal of Agricultural Marketing* and *Indian Journal of Agricultural Economics*. He was conferred the IDRC India Social Science Research Award for his work on the public distribution system. His areas of specialization include agricultural economics, rural development and development economics.

The Quality Brew

Productivity and Marketing of Coffee in India

Gana Shruthy M.K. and Parmod Kumar

Routledge
Taylor & Francis Group
LONDON AND NEW YORK

First published 2025
by Routledge
4 Park Square, Milton Park, Abingdon, Oxon OX14 4RN

and by Routledge
605 Third Avenue, New York, NY 10158

Routledge is an imprint of the Taylor & Francis Group, an informa business

British Library Cataloguing-in-Publication Data
A catalogue record for this book is available from the British Library

ISBN: 978-1-032-44007-1 (hbk)
ISBN: 978-1-032-89205-4 (pbk)
ISBN: 978-1-003-54172-1 (ebk)

DOI: 10.4324/9781003541721

Typeset in Sabon
by Deanta Global Publishing Services, Chennai, India

Dedicated to the parents of the first author
Who devoted their entire life growing coffee crop, putting all their love and affection and nurturing the coffee crop like their own children...

Contents

Figures

Tables

Preface

Coffee is a major export earner wherein there exists smallholders' domination both in terms of area and production. Given the bearing of coffee on the livelihoods of small growers and workers engaged therein, the Coffee Board has been undertaking a number of measures for building its international competitiveness and enhancing the economic viability of the growers. These initiatives included, but not limited to, the establishment of R&D institutions and an elaborate extension system to help increasing productivity inter alia by developing new coffee varieties and new cultural practices. Varied initiatives notwithstanding, the smallholder-dominated coffee sector is said to be under unprecedented crisis conditions attributable to a host of factors like declining productivity, incidence of new pests and diseases, adverse price trends in the case of Arabica, the preferred variety, and, above all, issues related to global warming-induced climate change. In this context, this book deals comprehensively with almost all the pertinent issues with respect to the production and marketing of coffee in India in general and Karnataka in particular. The study is based on a primary survey of planters and other functionaries in Kodagu and makes use of all the available secondary data to explore the relevant issues taken up for drawing inferences and to test the set hypotheses in the study.

There are several factors influencing the production of coffee directly or indirectly. An evidence-based understanding of these factors is crucial for identifying the most enhancing and limiting factors influencing its productivity. An efficient recording and analysis of data is required to establish credible inferences in this regard. It was, therefore, worth to carry out scientific research on local issues to develop alternative solutions to the problems under investigation. This study is intended to provide planters, policymakers, administrators, stakeholders, researchers and students with a fresh perspective on factors such as age of coffee trees, elevation of estates, estate-level annual and critical rains, shade trees, infestation of White Stem Borer (WSB), holding size and so on.

At the macrolevel, the study assumes importance in the current context of ongoing nationwide discussion on how to enhance productivity and improve coffee production; better avenues for coffee marketing; national and international developments in coffee such as certification issues, sustainable and fair trade coffee production, the concept of organic coffee and so on. Besides the issues of production, the book details issues relating to prices and marketing of coffee. It begins with analyzing monthly price variations across time periods and co-integration between local markets and foreign markets, using time series econometric tools. The study also analyzes different marketing channels, marketing margins and marketing efficiency. Further, it deals with various issues confronted by growers, traders and exporters with respect to production, processing, marketing and trade. Finally, a detailed analysis of the role of institutions in the production, marketing and export of coffee is provided at length with (a) the land tenure system, (b) new coffee bill of 2016 – in comparison with the Coffee Act of 1942 and (c) the role of the Coffee Board during the pre- and post-liberalization period in production, extension, knowledge development, export and sustainable coffee production. Finally, there is a detailed discussion on IT applications along with online trading and the role of various associations involved in the coffee sector. With this backdrop, the book unfolds major factors and institutions influencing productivity, marketing and trade in coffee of both Arabica and Robusta varieties.

The main inferences in the book are based on primary data, both qualitative (in-depth interviews, extensive field observations and review of documents) and quantitative data collected from 60 Arabica and 60 Robusta growers in Kodagu, classified as small (<2 ha), semi-medium (2–10 ha), medium (10–25 ha) and large growers (>25 ha). The primary data were collected during the period of March–August 2016. In order to understand the major factors influencing the productivity of coffee, seven major factors were identified, viz. age of coffee trees (years), elevation of estates (m.a.s.l), average annual rainfall (mm), rainfall during the critical period (mm), area under coffee (ha), number of shade trees per ha and WSB incidence (uprooted Arabica plants per ha). The influence of these factors on coffee productivity has been analyzed in depth using various tools of econometrics.

The results reveal the existence of a strong inverse U-shaped relation between age and productivity in both the varieties of coffee. The productivity of coffee initially increased with age, reached a maximum and further declined with increasing age. The inflection points in the case of Arabica and Robusta were 24 and 30 years, respectively, wherein the maximum productivity was observed, and thereafter productivity started declining. Although it is well known that age has a strong bearing on productivity, the phenomenon has not adequately been covered in prior literature. The study is the first of its kind to highlight the relation between the age of coffee trees and productivity. The non-availability of secondary sources of data remains one of the reasons for the dearth of research in the context of age. In addition to age, this study also documents productivity differences at different elevations. Another important inference drawn in this study is the infestation of WSB and the radical shift from Arabica to Robusta and the influence of genetic, environmental and management factors on WSB incidences.

With respect to the elevation of coffee estates, the average elevations among the surveyed Arabica and Robusta estates were 947 m.a.s.l. and 924 m.a.s.l., respectively. The productivity of Arabica increased with an increase in elevation. In contrast, the Robusta estates exhibited higher productivity at lower altitudes. Regarding rainfall, the Robusta estates significantly received heavier rainfall (2578 mm) than Arabica estates (1866 mm). In Arabica, the optimum productivity of 1000 kg/ha was found in estates receiving annual rainfall between 1651 mm and 2286 mm. However, in Robusta, the optimum productivity of 1600 kg/ha was found between 1058 and 2116 mm annual rainfall. Besides, a certain amount of blossom and backing showers was required during critical periods. During the study period, the coffee growers received timely and adequate blossom and backing showers that impacted its productivity. Overall, there existed a linear relation between critical rains and the productivity of coffee.

The next factor taken into account was the number of shade trees per ha. The Arabica estates maintained a higher shade density (237 shade trees per ha) compared to Robusta (147 shade trees per ha). The results revealed that, in the case of Arabica, lower productivity (792 kg/ha) was associated with a lesser number of shade trees, viz. <200 shade trees per ha. The optimum productivity of Arabica was observed in estates wherein the number of shade trees ranged between 300 and 400 per ha. In contrast, higher productivity of Robusta (1630 kg/ha) was observed with a lower number of shade trees per ha (<200 shade trees). The study also analyzed the relation between farm size and productivity of coffee. The average area under coffee was found to be 18.16 ha and 11.57 ha for Arabica and Robusta, respectively. The maximum productivity in Arabica was observed between 26 ha and 50 ha of land holdings. In the case of Robusta, productivity increased with the size of holdings, reaching a maximum between 21 ha and 50 ha, and thereafter declined with the landholding size.

Finally, the study takes into account the incidence of White Stem Borer (WSB) in Arabica in terms of the number of Arabica trees uprooted per ha. The regression analysis indicated that the WSB had a strong adverse impact on Arabica productivity. Thus, WSB remained a major constraint for the productivity of Arabica. Further, the study investigated several factors (area under coffee, elevation, actual rainfall, shade trees, expenditure on pesticides, age of coffee trees) influencing the borer incidence. The overall incidence of WSB remained low at higher altitudes. The prevalence of acute borer incidence occurred in estates located at lower altitudes. Likewise, the uprooting of Arabica trees reduced with an increase in the number of shade trees such as Dadap (*Erythrina lithosperma*), Indian rosewood (*Dalbergia latifolia*), Jackfruit (*Artocarpus integrifolia*) and Mahogany (*Cedrela toona*). The expenses on pesticides such as chlorpyrifos, lindane and contof reduced the incidence of WSB in individual estates. The incidence with respect to age revealed that the borer severely attacked Arabica trees between 20 and 30 years of age, while older trees above 30 years of age suffered less incidence of borer attack. The cultural and chemical costs for borer control varied between ₹670–₹7500 and ₹7425–₹9000 per ha, respectively.

About 31 per cent of the Arabica and 52 per cent of the Robusta growers sold their produce at farm-gate, followed by depots, curing works, hullers, companies, online sales and direct exports. The marketing efficiency was lowest when sold at farm-gate, with Arabica at 0.37 and Robusta at 0.91. The quality parameters, such as moisture content of the beans and outturn percent, significantly influenced the net prices of growers. Apart from this, the area under coffee (ha) and harvesting costs (₹/ha) positively influenced the net prices for Robusta growers.

Finally, the study documents the role of institutions in the production and marketing of coffee. First, it takes into account the land tenure system as an institution in Kodagu. The productivity of Arabica was higher (813 kg/ha) in unredeemed land tenure, in contrast to the Robusta variety, wherein productivity was higher (1715 kg/ha) in redeemed land tenure. In both Arabica and Robusta varieties, the indigenous shade tree density was higher in the unredeemed land tenure system. A higher percentage of exotic silver oaks (*Grevillea robusta*) was found in redeemed land owing to the tree rights conferred on them, resulting in a major shift from traditional poly-cultures of native shade canopies to the silver-oak monoculture. Then, the study details the 'New Coffee Bill-2016' and its similarities with the Coffee Act, 1942. Regarding sustainable coffee production, the study documents the role of organic, fair-trade coffee, Rainforest Alliance, Utz certified and Common Codes for the Coffee Commumity (4C). The role of Information Technology (IT) and CAFE Movel extension services and major growers' associations is documented in this book.

The book concludes with major findings and various policy suggestions for improving the overall coffee production and marketing system. The study suggests the need for in-sensitizing the coffee growers to replant and rejuvenate the old, senile and diseased estates; second, to set a limit of cultivation of Arabica only at higher elevation belts; third, to conduct combined and coordinated inter-country research on WSB in Arabica, rather than isolated research, and developing pest/disease-resistant varieties through a combination of antixenosis and antibiosis characteristics; and last, creating niche markets for sustainable certified coffee, aiming at the sustainable development of the coffee sector in India.

Authors
04.09.2023

Foreword

Coffee is one of the plantation crops in India, with Karnataka playing a predominant role. With a total area of 2.44 lakh hectares and producing 2.22 lakh metric tonnes of coffee, Karnataka accounts for 54.5 per cent of the area and 71 per cent of the output in the country. Within Karnataka, Kodagu district is rightly considered the 'coffee capital', accounting for over 53 per cent of the production within the state. Coffee is also a major export earner wherein there exists smallholders' domination both in terms of area and production. Given the importance of coffee as an export earner coupled with its bearing on the livelihoods of small growers and workers engaged therein, the Ministry of Commerce, through the Coffee Board, has been undertaking a number of measures to build its international competitiveness and enhance the economic viability of the growers. These initiatives include, but not limited to, the establishment of R&D institutions and an elaborate extension system to help increasing productivity inter alia by developing new coffee varieties and new cultural practices. Varied initiatives notwithstanding, the smallholder-dominated coffee sector is said to be under unprecedented crisis conditions attributable to a host of factors like declining productivity, incidence of new pests and diseases, adverse price trends in the case of Arabica, the preferred variety and, above all, the issues related to global warming-induced climate change. In this context, this book, organized into nine chapters, deals comprehensively with almost all the pertinent issues with respect to the production and marketing of coffee in India. The study is based on a primary survey of 120 growers from Kodagu and made use of all the available secondary data to explore the relevant issues taken up for analysis.

The first three chapters set the background for the study by reviewing the existing studies, locating the research problem, discussing the research methodology and eminently familiarizing the readers with the socio-economic and ecological dimensions of coffee production in India. Chapter 4 examines factors affecting the productivity of coffee – both Arabica and Robusta – and analyzes the role of different factors like age, elevation, size of holdings, annual rainfall, critical rain, shade trees and pruning. Using regression analysis, the study has come out with certain conclusions, such as an inverted

U-shaped relationship between age and productivity, highlighting the need for timely replanting to ensure high productivity.

Chapter 5 analyzes the impact of the coffee white stem borer on the productivity of the Arabica variety of coffee. The issue has assumed significance in the context of declining productivity of Arabica in the recent past, which in turn has induced growers to substitute the Robusta variety for Arabica. Using the regression model, the study analyzes the factors influencing the uprooting of Arabica per hectare. The estimated regression model highlights the bearing of factors like elevation, rainfall, shade trees per hectare, expenditure on pesticides and age structure. The study also examines the effectiveness of the institutional interventions by the Coffee Board to address this issue.

Cost and returns from the cultivation of coffee are analyzed in Chapter 6. Given the current level of productivity and prevailing market prices, returns over variable cost were found to be much higher for Robusta as compared to Arabica. The return per rupee invested was higher for Robusta (1.28) as compared to Arabica (0.99). Chapter 7 deals with a number of issues relating to prices and the market. It begins with an analysis of the price variation across different months and co-integration between local markets and foreign markets, using time series econometric tools. The study also analyzed different marketing channels, marketing margin and marketing efficiency. The overall marketing efficiency was found to be higher in case of Arabica parchment as compared to Robusta cherry. Finally, the chapter dealt with various issues confronted by growers, traders and exporters with respect to production, processing, marketing and trade.

Chapter 8 presents a detailed analysis of the role of institutions in the production, marketing and export of coffee. The chapter deals at length with (a) the land tenure system, (b) new coffee bill of 2016 – in comparison with the coffee act of 1942, and (c) the role of the Coffee Board during the pre- and post-liberalization periods in production, extension, knowledge development, export and sustainable coffee production. Finally, there is a detailed discussion on IT applications along with online trading and the role of various associations involved in the coffee sector.

On the whole, the book presents a fairly comprehensive analysis of the present state of the coffee sector with respect to varied aspects of production and marketing and has significantly contributed to our understanding of this sector. The conclusions drawn are of much relevance for both theory and policy, not only for the coffee sector but also for other crops in the plantation sector. Hence, I consider it a must-read for researchers, policymakers and all those interested in plantation crops, in general, and coffee, in particular.

Professor K.J. Joseph
Director, Gulati Institute of Finance and Taxation (GIFT)
Trivandrum. Formerly Ministry of Commerce Chair
Centre for Development Studies
Thiruvananthapuram, Kerala, India

Chapter 1

Introduction

1.1 Introduction

Coffee is a globally traded commodity with high demand in world trade. The developing nations contribute 90 per cent of coffee production, while 70 per cent of it is consumed only in industrialized countries (Ponte, 2004). About 125 million people across the globe are dependent on this sector for their livelihoods, which also contributes to three-fourths of total export earnings in certain developing nations (Dubois, 2001). Thus, coffee is a vital commodity of trade in developing economies. India, for instance, earned foreign exchange worth ₹7655.50 crore from the exports of coffee in 2021–22 (Indian Coffee Board, 2023).

Coffee is one of the most traded agricultural commodities in the world. In 2017, 70 per cent of total coffee production, worth US$ 19 billion,[1] was exported, while the coffee market size was valued at US$ 116 billion in 2021. In 2017, the sector had a retail market value of US$ 83 billion and provided jobs for 125 million people.[2] Coffee is grown on 12.5 million farms worldwide, with around two-thirds to three-fourths being smallholder farms primarily located in developing countries, including 22 Less Developed Countries (LDCs).[3,4] Brazil is the world's largest producer of premium Arabica coffee, while Vietnam is the world's largest producer of Robusta coffee. The largest coffee-producing and -exporting countries in 2017 were Brazil (US$ 4.6 billion), Vietnam (US$ 3.5 billion) and Colombia (US$ 2.58 billion), while the largest importing countries during the same period were the United States (US$ 6.3 billion), Germany (US$ 3.5 billion) and France (US$ 2.8 billion).[5] The global supply–demand balance for coffee varies from year to year; for example, in 2016 it closed with a supply deficit. Coffee supply growth outpaced demand growth in 2017 at rates of 5 per cent versus 2 per cent, respectively, resulting in a global coffee surplus of around 250,000 metric tonnes, with a surplus also estimated for 2018.[6] Coffee has gained popularity among the young population in the last decade, with substantial disposable income and a higher standard of living pushing premium coffee consumption. The sector is projected to grow, fuelled by increasing demand

DOI: 10.4324/9781003541721-1

not only from importing countries but also from producing countries. In emerging economies that have not traditionally been among the major coffee importers, such as Brazil, Indonesia and China, there is an expansion of retail options and coffee-based products such as ready-to-drink products.[7]

The coffee market is booming as consumption increases across all age groups, and global coffee exports reached a total of 10.61 million bags in 2021, indicating great demand worldwide. Especially in India, China and the Philippines, the popularity of coffee among the younger generation is increasing, and the demand for coffee beans is increasing. The awareness of the health benefits of coffee, such as reducing the risk of diabetes, fat burning and high energy intake from caffeine spikes, is further increasing this demand. The growth of local and global franchise coffee shops is driving the demand for a variety of coffee, as customers are willing to pay for high-quality coffee, which has become a medium for socializing. As a result, coffee shops such as Starbucks, Costa and Café Coffee Day (CCDs) are booming. With the growing cafe culture of today's younger generation and rising disposable income, employee demographics, urbanization and gastronomy are some of the factors contributing to the growth of the global coffee market. Changes in labour culture, especially in the corporate industry, are expected to further stimulate demand in the global market as well as improve living standards. Companies that are revising their service strategies and modern brands to increase customer satisfaction and loyalty are other factors that are improving the global coffee market. Increasing demand for organic varieties is expected to further drive demand for coffee shortly due to increased awareness of the health benefits of coffee. With the introduction of various flavours by baristas, CCDs and Starbucks, the customer experience scenario has been enhanced.

1.2 Global coffee market regional insights[8]

The European region dominates the global coffee market with a consumption of 54,065 million coffee bags per 60 kg. Europe accounted for 35 per cent of the world market share. The high demand for coffee in the European region and a large number of consumers of speciality coffee mean that 86 per cent of green coffee is imported directly from the country of origin. The Asia-Pacific region is the second-largest region with 2900 tonnes of coffee consumption due to increased coffee consumption in China, India and the Philippines, where coffee demand is highest. The Asia-Pacific region shows the potential to overtake Europe in the next five years, driven by increased demand for coffee as Southeast Asian countries emerge as more demand intensive countries by 2020. North America accounted for 4860 tonnes of coffee consumption, while South America accounted for 1630 tonnes. Coffee demand in North and South America has been constant over the past four years. The key players in the global coffee market are Steamy Beans Coffee,

The Busy Bean, Boston Barista, Jumpstart Coffee, Mugs Coffee, The Steam Room, Espresso Express, Nestle, JM Smucker Company, The Kraft Heinz Company, Starbucks Corporation, Keurig Dr. Pepper, F. Gavina and Sons, Inc., JDE Peet's, AMT Coffee Ltd, Lavazza and Trung Nguyen.

1.3 Coffee production in India

Coffee was introduced into India in 1600 AD by a Muslim saint named Baba Budan, who smuggled seven seeds of coffee from Yemen and sowed them in the Chandragiri hills of Chikmagalur district (Ukers, 1935). In 1820, the British, realizing the commercial significance of coffee, established Arabica coffee plantations on a large scale in South India. The period between 1825 and 1869 marked a golden era for coffee production in India. In 1869, it was estimated that about 120,000 acres of coffee plantations were established in South India (Joy, 2004). However, the same year also witnessed a major outbreak of pests and diseases like White Stem Borer[9] (*Xylotrechus quadripes*), Coffee Leaf Rust (*Hemileia vastatrix*) and green bug (*Coccus viridis*) that was witnessed for the first time in India, posing a huge threat to the Indian coffee industry (Ambinakudige, 2006). To overcome the problems faced with the Arabica variety of coffee, Robusta coffee (*Coffea canephora*) was introduced into India from Indonesia and Sri Lanka by the end of the 19th century.

Botanically, coffee is a member of the Rubiaceae family. There are two commercially cultivated species of coffee in India, viz. *Coffea arabica* (Arabica coffee) and *Coffea canephora* (Robusta coffee). *Coffea arabica* contains a lesser caffeine content, with a milder, juicier content, which is considered to be of higher quality than Robusta, but it is highly susceptible to various pests and diseases, viz. White Stem Borer[1] (*Xylotrechus quadripes*) and Coffee Leaf Rust disease caused by *Hemileia vastatrix* (Willson, 1999). *Coffea canephora* has a higher caffeine content, is resistant to pests and diseases and requires fewer management practices. *Coffea canephora* and *Coffea arabica*, respectively, account for 72 per cent and 28 per cent of total coffee production in India (Indian Coffee Board, 2023).

Coffee is a perennial crop. The average economic bearing period of coffee is 30–40 years, though some plantations bear up to 100 years (Duke, 1983). After planting, the initial four years of the gestation period are the establishment period. The coffee shrubs start bearing only after 3–4 years after planting. To obtain economic returns, it takes about 8 years for Arabica. The Robusta plantations require 10–12 years for economic yielding under good management practices like pruning, irrigation, shade regulation, fertilizing and pest and disease management (Titus, 2011).

In India, Arabica coffee is predominantly processed by the wet method to form parchment, while Robusta coffee is normally dry processed. However, about 15–20 per cent of Indian Robusta is also wet processed to form parchments, which commands a higher premium of US$ 300 per metric tonne

(Achoth, 2005). The prices of cherry and parchment coffee vary invariably. The average farm-gate price for Robusta cherry and Robusta parchment was ₹4430 per bag[10] and ₹8601 per bag, respectively, while Arabica cherry and Arabica parchment commanded a price of ₹8297 per bag and ₹15,782 per bag, respectively, in 2021–22 (Indian Coffee Board, 2023).

Prior to liberalization, coffee was the single product that was marketed through a statutory organization, viz. Coffee Board, that controlled the coffee industry, whose functions extended from collecting coffee from producers to processing and marketing of coffee. The monopoly of the Coffee Board was stopped, and coffee market was fully liberalized in 1996. Post liberalization, the functions of the Coffee Board were limited only to research & development, quality checks, market intelligence & information dissemination, market promotion and extension activities. Consequent to the dismantling of the Coffee Board, private traders and intermediaries took over the functions of coffee marketing.

1.4 Statement of the problem

The cultivation of coffee in India is traditionally confined to the hilly tracts of the Western Ghats (Velmourougane et al., 2004). India's share in total global coffee production was 3.41 per cent and its share in global exports was 5.11 per cent in 2021–22. In India, Karnataka stands first in area, production and productivity. The area under coffee cultivation in Karnataka was 2.46 lakh hectares in 2021–22, accounting for 54.47 per cent of total country's area under coffee. The state contributes 70.60 per cent of the total country's production, with the production of about 2,35,500metric tonnes of coffee in 2020–21, out of which 1,62,500 metric tonnes accounted for Robusta and 73,000 metric tonnes of Arabica. Tables 1.1a–1.5a present the area, production and productivity of coffee in India and its major states, while Table 1.6a presents the share of Kodagu district (our study area) in the state of Karnataka during the period from 2008–09 to 2020–21.

Table 1.1a indicates the area, production and productivity of coffee in India over the years. On average, Arabica and Robusta comprised about 47.23 and 52.77 per cent, respectively, of total area under coffee cultivation. However, the production of Arabica was 2.26 times lower than that of Robusta. The average productivity of Arabica over the years was merely 523 kg/ha, while that of Robusta was 1052 kg/ha. In general, statistics in these tables reveal interesting trends. While the area under Arabica increased by 2.5 per cent per annum over the period of 2008–09 to 2020–21, its productivity declined by around 2.1 per cent per annum during the same time period. In contrast, both the area and productivity of Robusta showed positive increases by 1 per cent and 0.37 per cent per annum, respectively, during the same time period. Consequently, despite a much larger increase in the area under Arabica compared to Robusta, the production of Arabica increased by only 0.37 per cent per annum compared to 1.38 per cent increase in the production of Robusta during the same time period of 2008–09 to 2020–21. The area and production trends in major states present even more interesting facts.

Table 1.2a indicates the overall area, production and productivity of coffee cultivation in Karnataka over the years. The overall area under Arabica and Robusta comprised 46.33 and 53.67 per cent of the total area under coffee cultivation, respectively. The statistics in the table indicates that the overall production of the state over the years was 2.22 lakh MT, with the production of Arabica and Robusta being 33.45 and 66.55 per cent, respectively. The overall productivity of Robusta was 1.71 times higher than Arabica in the state. Looking at the trends in area and production of the two varieties, unlike the increase in the area under Arabica at the all-India level, there was 0.07 per cent per annum decline in the area under Arabica in the state of Karnataka, whereas area under Robusta in Karnataka increased by 1.65 per cent per annum. However, the productivity of both Arabica and Robusta declined by less than 1 per cent and 0.1 per cent per annum against the rise of productivity in Robusta at the all-India level for the period mentioned earlier. As a result, the production of Arabica declined by 1.54 per cent per annum in Karnataka compared to 1.35 per cent per annum increase in the production of Robusta in Karnataka for the period of 2008–09 to 2020–21. On the one hand, there was a rapid decline in the area under Arabica from 110,298 ha in 2008–09 to 99,004 ha in 2020–21, while on the other hand, productivity also declined from 749 kg per ha in 2008–09 to 737 kg per ha in 2020–21 with a simultaneous increase in the area under Robusta from 116,602 ha to 128,098 ha during the same time period with an increase in productivity of Robusta from 1239 kg per ha to 1269 kg per ha during the same period. Thus, these trends broadly indicate that Arabica coffee is being gradually replaced by Robusta in the state of Karnataka.

Tables 1.3a and 1.4a present coffee trends in area, production and productivity in the neighbouring states, namely Kerala and Tamil Nadu, the other two major coffee-producing states. The share of Arabica and Robusta presented opposite trends in these two states. The share of Arabica in total coffee area and production in Kerala was less than 5 per cent and less than 3 per cent, respectively, while the rest of the area and production was contributed by Robusta. In contrast, in Tamil Nadu, Arabica contributed 82 per cent in area and 75 per cent in production, while the remaining contribution came from the Robusta variety.

The growth rate in area as well as productivity of both the varieties was positive in the state of Kerala (Table 1.3a). The area under Arabica increased by 1 per cent per annum, while productivity increased by 1.9 per cent per annum during the period of 2008–09 to 2020–21. The Robusta area was almost constant as growth in the area in Robusta was only 0.07 per cent per annum, while productivity increased by less than 1 per cent per annum during the same period. As a result, the production of Arabica increased by 2.6

per cent per annum in Kerala, while the production of Robusta increased by less than 1 per cent during the above-mentioned period. Similarly, the area under Arabica and Robusta increased by 1.6 and 1.1 per cent per annum, respectively, in Tamil Nadu during the said period, but the productivity of both these varieties declined in Tamil Nadu. The Arabica productivity declined by 1.1 per cent per annum, while Robusta declined by 0.74 per cent per annum. Consequently, the production was almost stagnant during this period as both these varieties observed –0.08 and –0.07 per cent per annum growth rate, respectively, for the period 2008–09 to 2020–21. Thus, it can be concluded that the overall production of coffee in India was not very encouraging during this more than a decade's period due to either a decline in area of Arabica or Robusta in one or the other state and more importantly that of productivity, which hardly saw a consistent increase in the major states where coffee is grown, namely Karnataka, Kerala and Tamil Nadu.

Finally, Table 1.5a presents a summary of coffee area and production in the three states of Karnataka, Kerala and Tamil Nadu. It is clear from the table that Karnataka occupies a 60 per cent share in the area under coffee, while it occupies a 72 per cent share in all India production. Kerala, on the other hand, contributes 22 per cent share in area and 21 per cent share in production, while Tamil Nadu's share is 8.5 per cent in area and only 5.7 per cent in production. Thus, these three states together contribute more than 90 per cent of the area under coffee and more than 98 per cent of the share in production, with only less than 10 per cent of the area and 2 per cent of the production contributed by the rest of India. Although Karnataka's area and production of Arabica is declining, the state still contributes around 60 per cent of the area of Arabica and slightly less than 80 per cent of production. Tamil Nadu is the next important state, contributing around 15 per cent of the area under Arabica and less than 14 per cent of the share in production. Kerala contributes relatively more in Robusta area and production, with a share of the All India area of slightly less than 40 per cent and a production share of slightly less than 30 per cent.

Table 1.6a indicates the area, production and productivity of coffee in Kodagu over the years. Over the years, the district comprised about 73.28 and 26.72 per cent, respectively, of the total area under Robusta and Arabica cultivation. The Robusta production was higher in the district (4.94 times higher than Arabica). The overall productivity of Arabica and Robusta over the years was 714 kg/ha and 1285 kg/ha, respectively. The Kodagu district, occupying the eastern slopes of the Western Ghats, is the largest coffee-growing district in the country (Navendu, 2009). The area of Kodagu is 4.10 lakh hectares, out of which 1.02 lakh hectares is under coffee cultivation. Out of the total area under coffee, about 75,199 hectares are under Robusta cultivation, while about 26,881 hectares of land is under Arabica cultivation. The district alone contributes 45.1 per cent of the state's share in area and 52.5 per cent of country's production. Kodagu accounted for 1,34,420 metric tonnes of coffee, out of which 20,500 MT was Arabica and 1,13,000 MT was Robusta (Indian Coffee Board, 2017).

Over the years, the country witnessed a structural shift from Arabica to Robusta coffee in India. The production of Arabica coffee decreased from 82 per cent in 1950–51 to merely 29 per cent in 2017–18. Besides, the national average productivity of Arabica[11] coffee was just 560 kg/ha, while that of Robusta was 1012 kg/ha. At the microlevel, the average productivity of Arabica in Kodagu was 634 kg/ha, much lower than the productivity of Robusta in the district, viz. 1367 kg/ha (Indian Coffee Board, 2017). Given these facts, it is interesting to know what are the factors that limit the productivity of Arabica and Robusta varieties of coffee in the district. A study on individual factors affecting the productivity of Arabica and Robusta coffee is essential to understand the productivity gaps existing in the region due to the most limiting factors. The study also assumes importance in the current context of the ongoing nationwide discussion on how to enhance productivity and boost production. This study would provide stakeholders with the necessary inputs to frame 'location-specific policies' to enhance productivity and further boost production as targeted by the Coffee Board.

Besides, coffee production in India is overwhelmingly a small growers' activity, who comprise 98.90 per cent of total land holdings. The small growers are at stake with uncertain fluctuations in prices. Additionally, the cost of cultivation of Indian coffee is increasing by 12–15 per cent every year and continues to rise (Business Standard, 2010), which subsequently affects the producer's income and profitability. With this rise in cost structure, will coffee production remain profitable in the future? As Chattopadhyay and John (2007) report, distress sales were widely rampant in coffee-growing regions, where growers were forced to sell their coffee at prices lower than the cost of production.

In addition to the above problems, coffee markets witnessed wider fluctuations in prices in the post liberalization period. The fluctuations in international prices are reflected in domestic prices as well, leaving the growers at the mercy of market conditions. Given these constraints, it comes as no surprise that the majority of growers focus on increasing quantity rather than improving quality. At this juncture, the fundamental question we must attempt to answer is the following: 'Is the growth in the coffee industry suffering signs of quality dilutions? If so, how and where are the dilutions taking place?' This question needs serious interrogation.

1.5 Theoretical framework

There are various factors (genotypes, altitude, rainfall, pests and diseases) affecting agricultural productivity either directly or indirectly (Ramaila et al., 2011). The earliest work on this aspect was carried out by John Von Liebig, who formulated 'Law of Minimum' in 1840. Liebig's 'Law of Minimum' states that agricultural productivity is limited by the scarcest factor used in production. The limiting or scarcest factor is the one that is available in the least supply in comparison to the requirement of the crop (e.g., labour, water, nutrients), which had greatest impact on productivity. Under such circumstances, the improvement of any other factor other than the 'scarce factor' does not improve productivity. It is only through the improvement of the 'scarce factor' or 'limiting factor' that higher productivity can be realized (Liebig, 1855).

The scarce factor or limited resources was invalidated by the neoliberal economists like Theodore Schultz, who asserted that the 'Law of Substitution' and 'technological innovations' could replace the scarce factor. That was to say that limiting factors or resources used in production, when replaced by technological innovations, could further enhance agricultural productivity. Schultz, in his 'theory of peasant rationality', viewed that the peasants followed traditional agricultural methods imbibed by their 'forefathers', which limited productivity. When traditional factors were being substituted by modern technologies (particularly improved seeds, fertilizers, pesticides and irrigation systems), agricultural productivity could be enhanced. He suggested that to make agriculture more profitable, a set of modern factors or inputs needed to be developed and supplied to farmers. Once such a new set of factors was introduced to farmers, "farmers would turn sand into gold" (Schultz, 1964).

The factors that affected production also affected trade. For instance, in the Ricardian 'theory of comparative advantage' (1917), labour was the only factor of production, and differences in labour productivity were the key justification for trade between two countries under the assumption of differences in natural resources, climate, soil and technology between the countries. While the 'factor endowment theory' put forth by Heckscher and Ohlin in 1919 argues that trade occurred due to differences in factor endowment between countries. The H-O model stated that a country would produce and export goods that were intensive in its abundant factors of production and import goods that make intensive of its scarce factor of production and that an economy with abundant factors gain, while scarce production factor will lose from trade. However, the factor that was abundant in one country could be scarce in another. The comparative advantage arose in a country due to differences in abundant or scarce factor endowments like natural resources, labour, capital, which provided an export advantage. In other words, countries export what they can easily, cheaply and abundantly produce. Specifically, the theory highlights that a nation's factor endowments were key determinants of comparative advantage, which resulted in trade. For example, Brazil had a comparative advantage in the export of coffee owing to its abundant natural resources like soil and climatic conditions, which were well suited for coffee production. Likewise, the United States and Canada export wheat because those countries were endowed with temperate-zone land most favourable for wheat production.

The movement of a product from the point of production to the marketplace involved a number of stages. Each and every stage in which exchange was performed incurred transaction costs (North, 1989). The transaction costs were the 'costs of running the economic system' (Arrow, 1970). The transaction costs were heavily influenced by the distance between the place of production and the place of marketing, along with physical infrastructure. Ronald Coase put forth 'transaction cost theory' in 1937, where he argued that any market transactions were not costless. The theory intended to understand how each agent decided to execute their activities in the marketing environment to reduce their transaction costs (Rodriguez, 2014). If transaction costs remained low, the traders preferred the spot market. On the other hand, if transaction costs were high, traders preferred contracting and vertical integration to reduce the costs (Ruben et al., 2007). The transaction costs involved ex-ante costs and ex-post costs. The ex-ante costs included negotiating and monitoring agreements like searching for potential exchange agents like local traders, wholesalers and retailers who offered better market prices, screening of potential agents and bargaining over terms of exchange. The ex-post costs included governance costs, maladaptive costs and monitoring compliance with any transactions (Willamson, 1985)

In lieu with 'transaction cost theory', the agency theory developed by Jensen and Meckling in 1976 was concerned with the reduction of ex-ante transaction costs in terms of principal and agent transactions. The agency theory or principal-agency theory involved principal and agent relationships (e.g., contractual relationships between farmer and a trader) and role of institutions in reducing transaction risks. The theory examined how a principal (owner of a coffee farm) could arrange contracts or agreements (selling contracts) with agents (local trader) to act in a way that could maximize the principal's goals and reduce the transaction costs involved (Kirsten, 2009).

Coffee as a commodity was subject to differentiation through differences in varieties, origins, processing, branding and so on, with its implications on market segmentation and price discrimination (Donnet, 2007). The 'new trade theory' developed by Paul Krugman emphasized product differentiation, market segmentation and price discrimination, similar to the theory of monopolistic competition. The new trade theory suggested that if a firm could produce a product that was distinct from others, it commanded a higher market power in setting the prices. In such cases, the transition of a commodity to a differentiated product would result in changes in the market structure from perfect competition to monopolistic markets, which subsequently resulted in differences in prices and market shares.

Although product differentiation and specialization led to an increase in trade, it also resulted in 'trade traps'. One such theory was 'immiserizing growth' theory propounded by Jagadish Bhagwati, which argued that specializing more in the production of goods for export could make the whole country worse off since exporting more of a commodity lowered its price in global markets, thereby causing revenues from exports to fall. The immiserizing growth was more pronounced in developing countries for agricultural commodities such as coffee, cocoa and tea due to demand for such commodities were price-inelastic and that production expansion would result in lower prices, revenue and profitability (Akiyama, 2001).

The pessimistic view of global trade was propounded by the dependency theory developed by Hans Singer and Raul Prebisch in 1970. According to this theory, global markets were characterized by two types of countries, viz. core countries and periphery countries. The periphery or poor nations supplied cheap natural resources, labor and markets to the core or wealthy nations. There lies a global division of wealth, wherein the core countries became wealthier at the cost of the former. The dependency theory was further modified into the 'world system theory' by Wallerstein in 1974. According to Wallerstein, a core nation was characterized by productivity dominance, which allowed them to produce commodities of higher price and cheaper quality in comparison to other countries, which was followed by trade dominance and financial dominance of core countries. It was noted that the world system theory was similar to the Marxist perspective, wherein developed capitalistic nations dominated poor nations through colonies, reflecting class inequality. Meanwhile, Andre Gunder Frank, the leading dependency theorist, argued that the traditional form of colonialism was giving way to neo-colonialism (colonialism practiced in a new form) wherein multi-national companies (MNCs)[12] of developed countries exploited less developed countries (similar to colonies) for cheap labour, raw materials and markets. Currently, the coffee industry conforms to dependency theory, where the dominance and exploitation of MNCs are realized, which controlled the export of more than half of the world's coffee production.

According to the Hymer–Kindleberger theory of structural market imperfection 1969, the market imperfections were the basis of foreign direct investment and the existence of multinational corporations (MNCs). The imperfections created in markets were due to product differentiation, branding, marketing skills, differences in technology and economies of scale. They noted that MNCs enjoyed 'monopolistic and oligopolistic' advantages in the host countries. The firms operated abroad by way of collusive agreements in host nations and were determined by the degree of imperfection and the comparative rate of returns (Yamin, 1991).

From the institutional perspective, coffee markets experienced huge changes post-liberalization. Before liberalization, the Coffee Board (state) had a monopoly over the entire coffee sector in India. Any monopoly of the state was concluded by Kenneth Arrow as follows: "the state might frequently have a special role to play in resource allocation because, by its nature, it had monopoly of coercive power and could be used to economize on transaction costs" (Arrow, 1970). In contrast, 'the public choice theory' popularized by James Buchanan offered a pessimistic view of the state. It viewed bureaucratic government officials as self-interested agents and that the state would favour certain groups to maximize its own utility and personal interest, thereby leading to inefficiency in the system. According to this theory, the state was not neutral in the economic process and 'optimal ignorance' was observed from the state (Cullis and Jones, 1987). The public choice theory pointed out that when the government does not achieve the desired effect, it resulted in government failure. The government failure happened at times when government intervention followed a more inefficient allocation of goods and resources. Similarly, prior to liberalization, the state-run Coffee Board made delayed payments and inefficiency in the marketing system set off the growers to demand market liberalization (Akiyama, 2001).

The liberalization in coffee markets took place in three phases. Firstly, Internal Sales Quota (ISQ) was introduced in 1992, wherein producers were permitted to directly sell up to 30 per cent of their produce in domestic markets, with the remaining 70 per cent being auctioned by the Coffee Board. Following this was Free Sales Quota (FSQ) in 1994, where producers could market up to 50 per cent either in the domestic or export market, while the remaining 50 per cent was auctioned by the Coffee Board. The markets were finally liberalized in 1996, with Extended Free Sale Quota, where the producers could directly market 100 per cent of their produce either in the domestic or export market (Ravindranath, 1999). The markets from then onwards supported the theory of the 'free market system', wherein any government interference or state-controlled market monopoly was broken while supporting private ownership (Akiyama, 2001).

The free-market system was even favoured by classical economists like Adam Smith in association with his 'laissez-faire' system, where he pointed out that the role of the state should be kept minimal (as in the case of coffee, role of the Coffee Board is minimal) and that the participants in the system were motivated by self-interest. Likewise, liberalists like Hayek and Milton Friedman, who opposed state interference in markets, favoured free markets (Karnik, 1997). However, liberalized free markets experienced a wide network of intermediaries, private traders and exporters who exploited the system similar to a capitalistic model, where inequalities in income favour the interest of dominant classes, either in the form of monopoly or market power (Vanek, 1970). Pierre-Joseph Proudhon opined that free markets and voluntary exchange cannot co-exist in a capitalist economy. So, trade liberalization does not imply that markets tend to approach perfect competition but rather approach imperfect competition (Vettas, 2006).

The liberalized coffee market was essentially a free market economy, wherein the economy was run by free market forces of demand and supply with a minimal role of government. Further, the system also introduced several intermediaries into the market, ranging from local agents, village-level consolidators, agent middlemen, traders, roasters, retailers and exporters (Subervie, 2008). However, there is still debate on how much the Coffee

Board had to interfere, as small growers continue to demand a vital role for the Coffee Board in the marketing of their coffee (Pavitha, 2005).

The above market reforms had also changed the role of institutions that existed earlier. One major factor that determined the rate of change was path dependence. The path dependency theory takes into account the changes in capacity of the institutions by analyzing the historic structure or, in the strongest form, that the historic hangover was inefficient (institutions in coffee prior to liberalization). However, institutions not only were the norms, rules and regulations that structure human behaviour but also coordinated transactions to reduce transaction costs (Williamson, 1985). Elinor Ostrom and other institutional analysts view institutions as the rule of game that facilitated, guided and constrained individual behaviours. Here, the institutions were commonly understood as codes of behaviour that potentially reduced uncertainty, mediated self-interest and facilitated collective action. Finally, the governance determined "*Who can do what to whom and on whose authority*" (McGinnis, 2011)

1.6 Literature review

1.6.1 Review on factors affecting perennial crops

Kumar and Sharma (2006) studied supply response in perennial crops, namely rubber, coffee and tea. The price and yield risks impact producers' decisions in establishing new plantations or allocating areas under new plantations. The yield functions for perennial crops were estimated with reference to one-year lagged price, rainfall patterns, technology and one year lagged yield. Technological improvements like improved clones in rubber and higher inputs had a positive response on yields, while the productivity of coffee positively responded to rainfall at blossom showers during February, March and April. In the short run, tea growers responded more quickly to prices and would thereby go for intensive plucking. In the long run, supply response function depended on higher use of inputs, actual rainfall, price and yield of competing crops.

Rutherford et al. (2006) studied the impact of pests and diseases on coffee productivity. The Coffee White Stem Borer was reported to cause losses in yield as high as 25 per cent, more severe in *Coffea arabica* grown at altitudes below 1700 metres. The mealy bug insect rapidly multiplies in dry weather and also causes considerable damage. 'Root-knot nematodes' disrupt the root functioning and absorption capacity of coffee trees to absorb water and nutrients. In Tanzania, root-knot nematodes reduced yield by 20 per cent. The Coffee Leaf Rust (CLR) was the most important disease in *Coffea arabica* grown at lower altitudes below 1500 metres, causing yield losses in excess of 75 per cent where incidence was severe. The Coffee Berry disease caused yield loss up to 30 per cent even after the application of fungicide. Coffee Wilt in Uganda caused considerable damage, wherein 14 million trees were destroyed.

Chethana et al. (2010) studied the influence of shade canopy on coffee yield. The coffee yield was low (6 quintals/acre) when cultivated under higher shade intensity (>70 shade intensity) and natural tree cover. Higher productivity (8 quintals/acre) was recorded in low shade conditions (<70 shade intensity). The productivity difference between high shades and low shades was 2.9 quintals per acre. It was estimated that planters growing coffee under low shade conditions earned a net gain of ₹10.40 per kilogram, while planters incurred a loss of ₹15.50 per kilogram when grown under high shade intensity. The study suggested that the marginal loss in productivity for high-density coffee growers should be compensated through price premiums.

Nsibirwa (2010) studied the emerging factors affecting coffee productivity in Uganda. The increasing urbanization and industrialization had reduced the area under coffee cultivation in Uganda. The shorter rain periods damaged the production of coffee by obstructing the trees from reaching the highest floration, as the rainfall directly controlled the floration and cherry maturation process. The scarcity of hired labour and the population growth in the land, diversionary farming enterprises, prolonged price depression, absence of youth in coffee production, small-sized land holdings, gender issues, lack of political support and dysfunctional coffee research system were the other factors responsible for lower coffee production in Uganda.

Krishnamurthy et al. (2011) studied the impact of rainfall and temperature on the productivity of black pepper in major pepper-growing regions of Kerala (Wayanad, Idukki and Cannanore districts), Karnataka (Coorg district) and Tamil Nadu (Coimbatore district). The rainfall showers in the months of December and January had a negative impact on pepper productivity, while rainfall in the months of April and May had a positive correlation with pepper productivity. The minimum temperature positively affected the productivity of pepper in high elevation zones (Wayanad, Coorg), while both maximum and minimum temperatures adversely impacted the productivity of pepper in the plains. Rather than rainfall and temperature, other factors such as prices, cost of cultivation, and pest and disease management also influenced production and productivity of pepper.

Dutta (2011) analyzed the effects of various factors impacting the productivity of tea (*Camellia sinensis*) in Assam in terms of leaf area index (LAI). The main factors impacting productivity in tea were age, nitrogen application and pruning. The studies have shown that age adversely influenced the productivity of tea. Younger bushes were associated with higher yields, while declining productivity was observed in older bushes. Though nitrogen fertilization was widely practised in tea estates, a negative impact of nitrogen application was noticed in older plants that resulted in higher nutrient imbalances and soil acidification. Good management practices like pruning positively improve plant productivity by stimulating the production of younger leaves and the development of younger shoots.

Watheri (2014) documented that the education level of growers was highly correlated to the productivity of coffee per tree. This could be due to better understanding and technological knowledge, while the higher income level also determined the affordability of inputs. On the other hand, lower productivity was also attributed to lower utilization of inputs. For instance, underutilization of fertilizers (application of 320 grams of fertilizers per year as against 550 grams per year), 0.8 debes of manure against the recommended 1 debe per plant per year, lower adoption of irrigation viz. adopted by less than 1 per cent of the growers, while only 60 per cent adopted chemical sprays.

Mesike et al. (2014) studied the influence of rainfall on the productivity of rubber, using time series data from 1971 to 2009. The results indicated an inverse relationship between rainfall and rubber productivity (–0.132). Heavy rainfall affects the tapping operations of rubber, thereby reducing the latex yield. The study thereby recommended defensive waterproof vessels for the collection of latex during the monsoons so as to block the latex from being washed away by rains.

Evizal et al. (2016) examined the diversity of shade trees on the productivity of coffee in Sumberjaya, West Lampung, Indonesia. In the West Lampung region, coffee was cultivated on individual land and on state land of the Community Forest Program (CFP) using several shade diversities comprising 36 species of shade trees. The diversity indexes used were, viz. Shannon and Simpson indexes. Higher diversity of shade trees was found in CFP coffee farms. The shade trees with a higher dominance index had a positive effect on the yield of coffee, and the per cent of Multi-Purpose Tree Species (MPTS) adversely influenced the productivity. The study notes that the type of land tenure, viz, Private or CFP, did not affect the productivity of coffee.

Ngeywo et al. (2015) assessed the influence of age, sex, marital status and land holdings on coffee production. The multi-stage and purposive random sampling method was used to obtain qualitative and quantitative data using a structured questionnaire from a sample of 227 growers. Using the Pearson correlation, the data was analyzed using descriptive statistics and inferential statistics at a 0.05 level of significance. The research findings showed that the mean production per tree was found 1.77 kg. The average age of coffee growers was 57 years. With regard to marital status, about 74.3 per cent were married. Gender, age and marital status had no influence on coffee production ($P > 0.05$), while the area under coffee had an influence on coffee production ($P = 0.023$).

1.6.2 Review on costs and returns of perennial crops

Reddy et al. (2003) estimated the cost of cultivation and cost of production of the Arabica variety of coffee in Chikmagalur. The average cost of cultivation was ₹52,955 per ha, while the cost of production was ₹51/kg. Labour costs, input costs and overhead expenditure accounted for 54 per cent, 30 per cent and 16 per cent of the total costs, respectively. The average yield of Arabica was estimated to be 1040 kg/ha. The gross return was found to be ₹57,200 per ha and the realized net profit was ₹4245 per ha at the price of ₹55 per kilogram of Arabica.

Devi et al. (2002) conducted an economic analysis of coffee in the Visakhapatnam district of Andhra Pradesh. The cost incurred for the establishment of coffee plantations worked out to be ₹23,656 per ha. The total costs incurred for the first year amounted to ₹13,090, which was 55.34 per cent of the total establishment cost. The establishment costs for the second year and third year amounted to ₹5327 per hectare. The total labour costs accounted for 56.44 per cent of the establishment costs, and material costs were 37.63 per cent of the total labour costs. The total maintenance costs of coffee plantations worked out to be ₹7609, ₹8129, ₹9243, ₹9405 and ₹10,326 per ha for the fourth, fifth, sixth, seventh and eighth years and subsequent years, respectively. The economic viability of coffee cultivation used both discounted and undiscounted measures. The net present value (NPV) was positive, while the benefit–cost (B:C) ratio was 1.68 and the internal rate of return was 23.85 per cent.

Umesh et al. (2005) studied costs and returns of Chintamani-1, Ullal-1, Ullal-2, Ullal-3 and and Ullal-4 varieties of cashew nuts in Karnataka. The establishment costs for all four varieties were ₹15,631/ha during the initial three years. The maintenance costs varied after the fourth year in all four varieties. The maintenance cost of Chintamani-1 varied between ₹5881– and 8254/ha, ₹5821 and 7229/ha in Ullal-2, ₹5812 and 7882/ha in Ullal-3 and ₹5640 and ₹8254/ha in Ullal-4. The net returns per hectare were highest in Chintamani-1, followed by Ullal-4, Ullal-3 and Ullal-2 in order of ₹61,314, ₹60,425, ₹49,672 and ₹34,231, respectively.

Prakash (2012) estimated the establishment cost of an areca nut garden, which worked out to be ₹6.38 lakh per ha in traditional areca growing regions (Malnad and Coastal), while ₹9 lakh per ha in non-traditional areca growing regions (Tungabhadra command areas). The cost of cultivation of the red boiled type (RBT) in traditional regions was ₹2.25 lakh per hectare, while in non-traditional regions it was ₹3.02 lakh per hectare. The cost of cultivation of the white chali type (WCT) of areca nut was ₹2.25 lakh per hectare. The average costs of production in RBT and WCT weres ₹13,470 per quintal and ₹10,123 per quintal, respectively. The growers incurred a net loss of ₹1905 per quintal of RBT and ₹2547 per quintal of WCT.

Saini et al. (2013) performed an economic analysis of guava production in Hisar, Fatehabad and Sirsa blocks in Haryana. The establishment cost was slightly higher in Fatehabad (₹41,250 per ha). This was followed by Sirsa (₹41,100 per ha) and Hisar (₹38,040 per ha). The average operational cost was ₹24,883 per ha. Guava starts bearing from the third year onwards at 24 quintals per ha up to a maximum of 210 quintals per ha in the seventh year. The net returns turned out to be negative for the three-year period. It gradually turned positive from ₹2269 per hectare in the fourth year to ₹71,947 per ha in the seventh year, after which the returns remained static for the next 25–30 years.

Mohammed et al. (2013) estimated the returns from coffee production in Kogi State in Nigeria, covering about 100 growers. The study found that coffee was profitable in the study region, with profitability index of 0.29. The estimated net farm income was N 8850.40 per household per hectare. The regression analysis of gender, age, household size, farm size and experience on the income of the growers indicated that these factors had a significant and positive impact on the net income of the growers. The profit level was affected by labour cost, processing, low yield and unavailability of market access. The study recommended that land ownership in the study region be re-visited, regulation of costs and educating the growers on fertilizer utilization.

Chengappa et al. (2014) estimated the economics of establishing a butterfly garden in existing coffee estates by adopting a partial budget approach. The estimated costs for establishing a butterfly garden were ₹64,000 per acre. The annual maintenance costs worked out to be ₹4280 per acre. To compensate for additional costs incurred and to maintain the same level of net returns, an additional premium of ₹167 per kg of Arabica parchment and ₹126 per kg of Robusta cherry was suggested. The net returns of ₹15,745 per acre would be realized in establishing a butterfly garden.

Narayana (2014) worked out the costs, returns and net economic gains to control Coffee Leaf Rust in Arabica by using sample survey data from 575 households in Karnataka, Kerala and Tamil Nadu. This was calculated by adopters of three recommended chemical controls, viz. first category for two rounds of Bordeaux mixture, second category for two rounds of systemic fungicides (Contaf) and third category for two rounds of systemic fungicides + one round of Bordeaux mixture. At the all-India level, the costs worked out to be ₹1842 per acre, ₹1956 per acre and ₹2872 per acre, respectively, for the above-mentioned categories. The gross returns were ₹41,666, ₹40,599 and ₹44,376 per acre for the first, second and third categories, respectively. The net economic gains as a percentage of gross returns worked out to be 35.34 per cent, 33.36 per cent and 36.97 per cent, respectively, for the three categories.

Swamy et al. (2016) performed an economic analysis of coffee cultivation in the Kodagu district in Karnataka state. Coffee productivity was found to be highest in the medium plantations, viz. 1051.20 kg per acre, followed by small (758.50 kg per acre) and large plantations (789.20 kg per acre). The average cost of cultivation was ₹17,041 per acre. The annual net return per acre was highest among the medium growers (₹26,109 per acre), followed by small (₹20,566.70 per acre) and large growers (₹18,572 per acre). The cost of production was highest among the large (₹22.50 per kg), followed by medium (₹21.20 per kg) and small growers (₹18.90 per kg).

1.6.3 Review on pests and diseases

Venugopal (1995) studied katte or mosaic viral disease in small cardamom (*Elettaria cardamomum Maton*) by collecting 134 disease samples from hot-spots of virus infections in South India and screening them in greenhouse, sick plot and hot spots. He estimated the yield losses to an extent of 68 per cent. The yield losses due to katte disease depended on the stage of crop infestation and duration of plant infestation. In the monocrop situation, the infestation in the first year reduced the crop yield to 38 per cent, followed by 62 per cent and 68.7 per cent in the second and third years, respectively, while the total decline of plants occurred after 3 to 5 years of infection.

Prabhakaran (1997) estimated yield loss in black pepper (*Piper nigrum*) due to several factors in Kannur using percentage avoidable loss (PAL). The average annual yield loss in pepper was found to be 33 per cent with the geometric mean estimator, while 34 per cent with the arithmetic mean estimator. The share of stand loss in the total losses was estimated to be about 13 per cent, while the remaining 20 per cent was due to a decline in productivity. The major stand losses were due to foot rot caused by *Phytophthora capsici* which contributed to 9 per cent losses. The annual average stand loss due to drought was found to be 4 per cent, while the annual losses due to slow wilt were relatively less. The pollu beetle (*Longitarsus nigripennis*) accounted for 13 per cent losses.

Oerke (2005) estimated the crop losses for wheat, rice, maize, potatoes, soybeans and cotton as well as various methods for pest control for the period of 2001–03 on a regional basis as well as at the global level. Among the crops, the total global potential losses due to pests varied from 50 per cent in wheat to more than 80 per cent in cotton production. However, the crop losses were found to be 26–29 per cent in soybeans, 31 per cent in wheat and 37 per cent in cotton. For maize, rice and potato it was found to be 40 per cent. The weeds accounted for the highest potential losses of 34 per cent. The animal pests and pathogens accounted for 18 and 16 per cent of the total crop losses, respectively. The efficiency of crop protection was higher in cash crops than in food crops. However, despite pesticide applications, the crop losses have not decreased over the last 40 years.

Guenthner et al. (2001) estimated yield losses in potato due to late blight caused by the fungus *Phytophthora infestans*. The Delphi survey was conducted in order to estimate the impact of late blight on potato yields. It was estimated that under a no-late blight scenario, the yield was 5 per cent higher and storage losses decreased by 17 per cent. The impact of late blight on the revenue was $210.70 million. The quality impact included rejections and price discounts that were estimated at 64 per cent. The growers effectively controlled late blight but with a higher cost. The estimated fungicide costs were $77.10 million. The total costs averaged $507 per hectare, excluding non-fungicide control practices.

Singh et al. (2007) documented the percentage of crop losses in the major crops of India. The crop losses due to rice stem borer (*Scirpophaga incertulas*) ranged from 10 to 48 per cent, leaf folder (*Cnaphalocrocis medinalis*) from 10 to 50 per cent, whorl maggot (*Hydrellia* spp.) from 20 to 30 per cent, gall midge (*Orseolia oryzae*) from 8 to 50 per cent and hispa (*Dicladispa armigera*) from 6 to 65 per cent. The army worm (*Mythimna separate*) in wheat accounted for 20–42 per cent of the crop losses and stem borer in sorghum (*Chilo partellus*) caused 55–83 per cent of the losses, whereas oriental army worm (*Mythimna separata*), shoot fly (*Atherigona soccata*) and earhead caterpillar (*Helicoverpa armigera*) caused 55.7 per cent, 22–80 per cent and 18–26 per cent of the losses, respectively.

Murphy et al. (2008) estimated the national average annual reduction in yield due to White Stem Borer in coffee. The losses were to the extent of 13.4 per cent and 14.80 per cent in Malawi and Zimbabwe, respectively. The losses converted to the production losses of 555 tonnes and US$745,705 in Malawi and 1007 tonnes and US$1,353,018 in Zimbabwe. In India, they estimated that the loss of 100 plants infested per ha could cause a revenue loss of US$10 million. The average yield loss was 0.80 kg/healthy tree. The study estimated that during 2002–03, the average yield in Karnataka was lower by 11 per cent. The yield loss percentages were 13 per cent, 10 per cent and 8.3 per cent in Hassan, Chikmagalur and Kodagu, respectively.

Fernandes et al. (2011) estimated the economic injury level (EIL) for the coffee berry borer by using attractive traps in Brazil with an objective to determine the economic injury level (EIL) considering the qualitative and quantitative losses for conventional and organic coffee. Considering the quantitative losses for conventional coffee, the economic injury level (EIL) ranged between 7.90 and 23.70 per cent, whereas for organic coffee, the EIL varied between 24.40 and 47.60 per cent. Taking into account both the qualitative and quantitative losses, the EIL was 4.3 per cent for both conventional and organic coffee. The studies found that the EILs for the berry borer associated with the coffee plant during the flowering, pinhead fruit and ripening stages were 426, 85 and 28 adults per trap, respectively.

Honger (2014) estimated total yield loss and quality loss in mango due to the disease. The percentage of fruit drop due to Anthracnose ranged between 26.60 and 32.30 per cent among the trees of different ages, averaging to 29.90 per cent. The fruit drop was observed from the beginning of the first week of fruit set. At the fruit maturity stage, the disease symptoms were found on the fruits hanging on the trees, resulting in poor quality. The disease incidence recorded from different plots ranged from 31.30 per cent to 32.80 per cent, while the disease severity index ranged from 3.20 to 3.30.

The percentage of fruits that reached maturity but could not be marketed alone accounted for 22.90 per cent of total fruit production.

1.6.4 Review on marketing

Gabre (2001) studied trader behaviour among grain wholesalers in Ethiopia. The study revealed that Ethiopian grain wholesalers were generally small-scale enterprises. The marketing costs like transportation costs and handling and other market activities represent about 83 per cent of gross margin, while traders' net margin was less than 5 per cent. The effective expansion of the grain market depended on the ability of traders to exchange grains anonymously with buyers and sellers at distant markets. The major difficulties faced by the traders include weak public information, lack of grain standardization, oral nature of the markets and limited legal enforcement of contracts. The grain market brokers performed multiple functions that included inspection and witnessing transactions, besides delivering 'public goods' such as market information and product classification as well play a pivotal role in price discovery.

Philip (2003) studied the marketing of spices with special reference to pepper and cardamom. The domestic spice traders face difficulty in procuring appropriate quality spice while they also face intense competition in the domestic market. The warehouse facilities are not readily available to small and medium traders and therefore they are unable to hold their product to take advantage of price hikes. There were no specific programmes from the Spices Board exclusively intended for domestic traders. The domestic factors such as competition, operation costs, value addition and sources of supply of spices were the most decisive factors in the trading of spices, while trading channels, financial constraints and factors like government facilities, warehousing facilities and stock velocity determined spice trading. The corporate exporters were able to secure more orders than partnership or sole traders, while exporters did not receive international prices for spices exported. This also included risks of contract defaults and settlement problems.

David et al. (2008) described two critical nodes in Indian coffee trade channels, i.e., agents and curing works, largely controlled by exporters. Exporters, being the most powerful actors, controlled the entire marketing chain. Conversely, the small growers were taken aside from marketing channels. The growers sell their coffee to local agents, who in turn transport to curing works, where secondary processing takes place. The cured coffee can be sold at the ICTA (Indian Coffee Traders Association) auctions held in Bangalore, sold to roasters or exported. The exporters sometimes own the curing works. Prior to export, cup testing is conducted in laboratories. Later on, the commodity is shipped for export. The major ports of shipment of Indian coffee are Cochin, Chennai, Mangalore and Tuticorn.

Chand (2010) estimated the factors influencing the participation of growers in gherkin contract farming. The results of the probit model indicated that age, farm size, net income and debt significantly influenced the participation of growers in contract farming. Age (0.05) and farm size (0.30) negatively influenced participation. In other words, as the age of the growers increased by 1 per cent, the probability of participation decreased by 0.05 per cent. Likewise, as the farm size increased, participation in contract farming reduced. If the indebtedness of the growers increased by 1 per cent, participation increased by 0.01 per cent.

Bhavya (2011) analyzed the value chain structure for coffee in Karnataka. The price spread in channel 1 (farmers-agents-curers-local roasters) was ₹35,350 per tonne of coffee. The roasters' selling price was ₹85,350 per tonnes of coffee, while the growers' selling price was ₹50,000 per tonne of coffee. The margins for agents, curers and roasters were found to be 3.34 per cent, 6.38 per cent and 15.25 per cent, respectively. However, the price spread in channel 2 (farmers-curers-exporters) was ₹28,720 per tonne of coffee. The exporters' selling price was ₹78,720, while the price received by the growers was ₹3200 per tonne of coffee. In this channel, the margins for curers and exporters were found to be 6.21 per cent and 13.94 per cent, respectively.

Hazarika (2011) analyzed the marketing problems of the tea industry in Assam. About 62.50 per cent of Indian tea was disposed of through public auctions annually. The participants in auctions were producers, brokers and warehouse keepers. The tea is stored in registered warehouses. The brokers draw samples for tasting and valuation. The brokers were powerful agents in auctioning the produce. The producers, however, do not have any active role in auctioning. The auction buyers were mostly big tea companies who contract with brokers to keep tea prices low. Therefore, growers were price takers in tea auctions. The actual market process occurred after auctioning, where processing, blending, packaging, advertising, wholesaling and retailing start. The profit margins were captured by MNCs with their enormous marketing skills. The producers were generally satisfied with the current auctioning system, as it was responsible for the smooth disposal of their produce.

Chengappa et al. (2014b) mapped the value chain for conventional, certified and organic coffee in Kodagu. The conventional chain comprised producers, village-level consolidators, traders, hullers, domestic companies, MNCs, curers, exporters, importers, domestic and international roasters, wholesalers, retailers and consumers The growers incurred a margin of ₹71,049, ₹121,420 and ₹82791 per tonne of clean coffee in the conventional, certified and organic Robusta value chains, respectively. In organic Robusta coffee, the domestic companies and retailers incurred a margin of ₹157,969 and ₹85,111 per ton. In the conventional chain, the planters, village-level consolidators, hullers, agents, curers, roasters and retailers' shares of the total margin were 43 per cent, 1.1 per cent, 6.3 per cent, 2 per cent, 13.90 per cent, 14.60 per cent and 18.60 per cent,respectively.

Ntimba (2017) assessed the marketing efficiency of Robusta coffee under the different supply channels in the Karagwe district of Tanzania. The analysis revealed that the coffee producers selling through village buyers were characterized by lower marketing efficiency (9.38) compared to the ones selling through private buyers (13.70) and cooperative societies (13.08). The study revealed that the coffee market concentration in the district measured 87.50 per cent and 90.30 per cent concentration ratios, and the Herfindahl–Hirschman Index (HHI) was about 0.29. These values establish a highly concentrated market with a few dominant traders who tend to exhibit oligopolistic behaviour, which impedes marketing price competition and thereby suppresses the bargaining power of growers.

Promme (2017) assessed the factors determining the type of rubber product the farmers were willing to sell and the factors influencing their market channel choice in Thailand. The primary data were collected from 280 rubber growers in the two southern provinces. The results of the logistic regression model revealed that the membership of farmer organization, access to knowledge and information and total area of immature rubber positively influenced the sale of fresh latex to the farmers.

1.6.5 Theoretical review on perennial models

There have been several models put forth by Bateman (1965), Ady (1968), Behrman (1968), Rajkrishna (1963) and Kumar and Sharma (2006) on the response of perennial crops to a number of factors of production. These major models have relied on the basic Nerlovian price expectation and acreage adjustment model, put forth by Nerlove in 1958. The Nerlovian price expectation model assumes that producers not only respond to the previous year's prices but also expected prices, given as

$A_t = a + bP_t^* + CA_{(t-1)} + U_t$
$(P_t^* - P^*_{(t-1)}) = \beta(P_{(t-1)} - P^*_{(t-1)})$, where
A_t = Actual acreage under crops in year t
P_t^* = Expected price of the crop in year t
P^*_{t-1} = Expected price of the crop in the preceding year
P_{t-1} = Actual price of the crop in the preceding year
U_t = Error term, a and b = constants
β = Acreage adjustment coefficient

Bateman (1965) had estimated the total output of cocoa by studying five cocoa growing regions (Volta, Eastern, Central, Western and Ashanti) of Ghana, owing to differences in soil conditions and production practices over the period of 1946–62. He made an assumption that prices influence the planting of cocoa, which in turn influences changes in output. In his perennial supply model, he emphasized the planting relationship and the output relationship.

$Qt^* = b_1 (\Sigma_{i=k}{}^{s-1} X_{t-i}) + b_2 (\Sigma^{\infty} X_{t-i})$
k = Age at which cocoa begins to yield
s = Year in which second phase of rapid increase in cocoa yield occurs
b_1 = Output per acre after first phase
b_2 = Output per acre after second phase

In his model he concluded that the yield of cocoa attains the first phase/plateau after 8 years. The maximum yields are achieved after 12 years. He considers prices lagged by 8 years and 12 years as independent variables. The model holds good, provided cocoa trees are not subjected to depreciation, i.e. yield remains stable after 12 years and does not decline. However, in reality, cocoa yield increases gradually till the peak, followed by a gradual reduction in yield until trees attain zero production. Moreover, he assumed an infinite life of the perennial crop, which was an unrealistic assumption.

Further, he considers rainfall and humidity factors as important variables for cocoa yield, which were expressed in first difference form. Rainfall during the months of March to June (formative period) is critical for cocoa crops. Coffee is a major substitute for cocoa in Ghana; therefore, deflated coffee prices lagged by 8 and 12 years were considered as independent variables. The first difference equation is given as follows:

$$\Delta Q_t = b_2 a_0 \beta + b_1 a_1 \beta P_{t-k} + (b_2 - b_1) a_1 \beta P_{t-s} + b_1 a_2 \beta C_t + (b_2 - b_1) a_2 \beta C_t + \\ C \Delta R_{t-1} + d \Delta H_{t-1} + e \Delta P_t + (1 - \beta)(\Delta Q_{t-1} - C \Delta R_{t-2} - d \Delta H_{t-2} - e \Delta P_{t-1}) + W_t$$

Q_t = Quantity of cocoa harvest in year 't'
b = Potential yield per acre
k = Age at which cocoa trees start to bear
R_{t-1} = Rainfall during the formative period between March and June
H_{t-1} = Humidity variable affecting yield
P_t = Producers' price in year 't'
C_t = Prices of coffee at time t
W_t = Error term

The above equation indicates that the changes in cocoa yield harvested each year are determined by factors such as the age of the tree, rainfall in critical stages of growth and humidity. The price of substitutes like coffee (C_t) indirectly determines the planting decisions of cocoa growers in the current year. Bateman's theory indicates that rainfall in the months of March–June has a positive impact on the yield of cocoa, while rainfall during July–August negatively impacts yield by stimulating the black pod virus.

In similar lines, Behrman (1968) adopted a partial adjustment approach for the estimation of supply response in cocoa. In his study on monopolistic cocoa pricing, he analyzed eight countries which included Ghana, Nigeria, Ivory Coast, Brazil, Ecuador, Venezuela, the Dominican Republic and the

Cameroon Republic for a period from 1947–48 to 1963–64. The Behrman equation is given in the following form:

$$\Delta Q_t = [(1-\delta) + (1-\beta)]\Delta Q_{t-1} - (1-\delta)(1-\beta)\Delta Q_{t-2} + \Delta Q_{t-2} + C\Delta P_t + [d - c(1-\delta)(1-\beta)]\Delta P_{t-1} + \{-d[(1-\delta) + (1-\beta)] + (1-\delta)(1-\beta)c\}\Delta P_{t-2} + (1-\delta)(1-\beta)d\Delta P_{t-3} + b_1\delta\beta[a_1\Delta P_{t-k} + a_2\Delta C_{t-k}] + (b_2 - b_1)\delta\beta[a_1\Delta P_{t-5} + a_2\Delta C_{t-s}] + U_t$$

where

Q_t = Quantity of cocoa harvest in year 't'
b = Potential yield per acre
k = Age at which cocoa trees start to bear
P_t = Price of cocoa in time 't'
C_t = Prices of coffee at time t
δ and β are expectation and adjustment coefficients, respectively

Wickens and Greenfield's (1973) model of perennial crops is an improvement of Nerlove and Behrman's model and comprises a Vintage production function, an investment production function and a harvesting function.

(a) Vintage production function is given as:

$$q^p_t = \Sigma^n d(i, t) I_{t-i}$$

where

q^p_t is the potential output.
$d(i,t)$ is the average productivity of the trees
I_{t-i} is the number of trees planted i years ago, survived in tth year
The Vintage production function indicates that potential production is a function of the average yield of trees and the number of trees in production.

(b) Investment function is given by

$$V = \Sigma^{\infty}_{i=0} (1+r)^{-t} \{(P^e_t - S^e_t)\, q^p_t - F_t - f_1(I_t)\}$$

where

V is the expected discounted net revenue
r is the rate of discount
P^e_t is the expected unit price
S^e_t is the unit cost of harvesting
F_t is fixed cost in year 't'
$f_1(I_t)$ is the planting cost
I_t is the number of trees in year t

The investment function indicates a long gestation lag in perennial crops, in which decision to invest in perennial crops is based on the expected discounted income stream from the investment.

(c) The harvesting function is given by

$$q_t = a_0 + a_1\, q^{P}{}_t + \Sigma^{m}{}_{i=0}\; biP_{t-I},$$

where

q_t is the actual production
$q^{P}{}_t$ is the potential output
bi is the harvesting response to price
m is the length of the lag

In the Indian context, Rajkrishna (1963) included rainfall, yield and total irrigated area in the Nerlovian model. He conducted studies in Punjab for major crops for the period of 1914–45. The following equation was formulated by Rajkrishna:

$$A_t = a_0 + b_0 P_{t-1} + c_0 Y_{t-1} + d_0 Z_{t-1} + e_0 R_t + g_0 A_{t-1} + V_t$$

A_t = Actual irrigated area
P = Post-harvest price
Y = Yield of the crop
Z = Total irrigated area of all crops of the season
R = Rainfall
V = Error term

The irrigated area is the dependent variable, while rainfall, yield and total irrigated area under all crops are taken as independent variables (non-price explanatory variables). He concluded that the net effect of the price factor can be properly measured only if non-price factors determining supply were well defined, with his findings that "the more correctly we are able to specify the non-price variable, the more significant the net regression coefficient and the elasticity of price variables". This means that non-price factors play a significant role in determining supply elasticity.

Kumar and Sharma (2006) estimated the perennial supply response function of coffee, tea and rubber. In the short run, the producers could adjust their supply within their existing productive capacity through intensive cultivation. The major determinants were the prevailing market price and technology that could shift production upward, while other crucial determinants were the age of the tree and rainfall. In the long run, producers could increase their yield by increasing area under cultivation and through replanting, expected profitability in the long run and expected profitability of substitutes. The equation is formulated as follows:

$$\mathrm{BA}^{*}_{it} - \mathrm{BA}_{it-1} = f(\mathrm{RP}_{it}^{e}, \mathrm{RP}_{jt}^{e}, Y_{it}^{e}, Y_{jt}^{e}, S_t^{e}, R_t, L_{t-1}, V_t)$$

where

BA^{*}_{it} = The Desired bearing area of the *i*th crop in year '*t*'
BA_{it-1} = The actual bearing area of the *i*th perennial crop in year $t - 1$
RP_{it}^{e} = The expected real price of the *i*th perennial crop
RP_{jt}^{e} = The expected real price of competing crop
Y_{it}^{e} = The expected yield of *i*th perennial crop
Y_{jt}^{e} = The expected yield of the *j*th perennial crop
S_{t}^{e} = The variable to account for changes in perceived risks
R_t = Rainfall index
L_{t-1} = Lagged dependent variable
V_t = Disturbance term
The new planting function is expressed as follows:
$NPA_{it} = f(RP_{it}^{e}, RP_{jt}^{e}, Y_{it}^{e}, Y_{jt}^{e}, S_{t}^{e}, L_{t-1}, R_t, V_t)$

The producers' decision on bringing new areas under cultivation depended on their own prices, prices of substitutes, yields of their own and competing crops, perceived risks and rainfall. In the short run, as prices of perennial crops increase, producers indulge in more intensive cultivation, whereas in the long run rainfall, yield, price risk factors and prices and yields of competing crops become major determinants of the planting area.

Chatterjee (2005) derived a consumption function for tea. The aggregate consumption of tea was given as 0, where M is the domestic per capita income and 'p' referred to vector prices, which included the price of own tea and prices of its substitutes and complements. The consumption of tea was inversely related to its own price and the price of complements, while it was positively related to the price of substitutes and income. Further, considering the impact of 'tastes and preferences' on tea consumption, the model was given as

$$C = \alpha_0 + \alpha_1 P_0 + \alpha_2 P_s + \alpha_3 Pc_1 + \alpha_4 P_{c2} + \alpha_5 M + \alpha_6 Trend + \varepsilon_2$$

where P_0 is the own price of tea
P_s is the price of substitutes
Pc_1 and Pc_2 are the prices of sugar and milk, respectively

The models stated above were the improvements from the Nerlovian supply response model, which proved to be much more beneficial for perennial crops like coffee, tea, cocoa, rubber and so on. These models can be used to study the factors that determine the production of perennial crops in Karnataka and Kerala.

1.7 Identified research gaps

- Factors responsible for productivity differences among different varieties of coffee.

- Per-unit realization of profit in Arabica and Robusta coffee and their associated cost structure.
- Quality aspects maintained by each and every actor in the market.
- Role of institutions in promoting the production and efficient marketing of coffee and protecting the interests of various actors in the chain.

1.8 Research questions

- What are the factors responsible for productivity differences in different varieties of coffee in Kodagu?
- How do the cost structure and profit differ in different varieties of coffee?
- Do producers, traders and exporters maintain quality while trading? If not, where and how are dilutions taking place?
- What is the role of institutions in supporting producers, traders and exporters?

1.9 Objectives of the study

- To study the major factors influencing the productivity of coffee in Kodagu.
- To study the economics of Arabica and Robusta varieties of coffee in the study region.
- To study the existing market structure and pattern followed in coffee marketing and analyze problems faced by producers, traders and exporters in maintaining quality
- To study the role institutions play in the process of coffee production, marketing and export.

1.10 Hypothesis

$H_{1.1}$: Coffee productivity makes an inverse U shape with respect to the age of coffee plantations in both Robusta and Arabica varieties.

$H_{1.2}$: Arabica productivity has a direct relationship with the elevation of estates, whereas in the case of Robusta, the relationship is inverse.

$H_{1.3}$: Productivity of both Arabica and Robusta has a positive relationship with respect to natural factors like rainfall (including critical rainfall) and shade canopy.

$H_{1.4}$: The White Stem Borer (WSB) affects productivity negatively only in the case of Arabica, while the Robusta variety is free from WSB incidence.

H_2: The Robusta variety of coffee offers higher returns on investment compared to the Arabica variety.

H_3: Marketing efficiency is least in Channel I (i.e., farm gate), while it improves in Channel II and Channel III (e.g., hullers, curers, coffee depots and so on).

H_4: There exists a strong and positive relationship between net price realization and the quality of coffee in terms of out-turn and moisture content.

1.11 Database and methodology

The study comprises both primary and secondary data, collected from 60 Arabica and 60 Robusta growers from the three tahsils of Kodagu. The primary data were also collected from the traders (5), hullers (4), curers (5), exporters (5) and roasters-cum-retailers (5). The secondary data pertaining to the study were collected mainly from publications of the Coffee Board of India and the International Coffee Organization (ICO). The detailed methodology is provided in Chapter 2.

1.12 An Overview

The book is divided into five sections: introduction, review of literature, database and methodology, results and discussion, and conclusions.

The introduction is provided in Chapter I, which comprises the statement of the problem and the theoretical framework of the study. The review of the literature on factors affecting perennial crops, costs and returns of perennial crops, pests and diseases, marketing of agricultural products and theoretical review of perennial models have been provided in this chapter. From the review of the literature, research gaps were identified, and research questions, objectives of the study and hypothesis were formulated accordingly.

The database and methodology used for the study are detailed in Chapter 2. The sources of data, sampling design, selection of study area, villages and respondents and market functionaries are provided in this chapter. Further, the general description of the study area, economic concepts and analytical tools and definition of terms and concepts are provided in this chapter.

The results and discussions are presented in five chapters, viz. from Chapter 3 to Chapter 8. Chapter 3 comprises the socio-economic and ecological dimensions of coffee production in Kodagu, detailing the socio-economic and estate profiles of the growers, intercrops in coffee estates and credit details of the growers.

The major factors influencing the productivity of Arabica and Robusta varieties of coffee in Kodagu are provided in Chapter 4. The age of coffee trees, elevation of estates, average annual rainfall, rainfall during critical periods, shade trees, area of coffee estates and their corresponding influence on productivity of coffee are provided in this chapter.

The coffee White Stem Borer (*Xylotrechus quadripes*) and its impact on the Arabica variety of coffee are detailed in Chapter 5. The opinion of Arabica growers about shifting towards the Robusta variety of coffee, factors responsible for borer attacks in Arabica, costs incurred for borer control and the institutional support for WSB control are covered in this chapter.

The economics of Arabica and Robusta cultivation in Kodagu are provided in Chapter 6. The total costs and revenue in coffee cultivation, returns from coffee cultivation, gross income and total resources used in coffee production are detailed in this chapter. However, the market structure and quality aspects of coffee are detailed in Chapter 7. The mode of sales of coffee, fluctuations in farm gate prices, sales of certified coffee, market integration, Johansen Co-integration, unit root test, price spread marketing efficiency of coffee, factors affecting net prices of coffee and problems faced by the producers in maintaining quality are provided in this chapter.

The role of institutions in production, marketing and exports of coffee are detailed in Chapter 8. The land tenure system as an institution in Kodagu, the New Coffee Bill 2016, the role of Coffee Board, the role of institutions in controlling major pests and diseases, the application of Information Technology (IT) in coffee production, online trading initiative, major associations in coffee processing and exports and major schemes for the promotion of exports are provided in this chapter.

The results of the primary survey along with discussions have been summarized and concluded in Chapter 9, followed by the reference section that includes all the literature cited in the thesis.

1.12 Limitations of the study

The following are the major limitations of the study:

The major limitation of the study is that there have been a number of other factors influencing the productivity of coffee such as temperature, genetics, soil conditions, changing weather and climate, migration of youth and other socio-economic and environmental factors. However, the current study does not analyze the impacts of all these factors.

i. The study is based on the primary data set, which is cross-sectional, collected at a specific point in time, and repeated surveys were not undertaken due to time and resource constraints.
ii. The study is confined to only one district, and findings of the study might be location-specific. A study spatially comparing other coffee-growing regions of the country would have strengthened the policy implications.
iii. The sample size of the growers is also considered a limiting factor, viz., only 60 Arabica and 60 Robusta growers. A larger sample size covering other coffee-growing districts of Karnataka may have resulted in deriving more conclusive statements.
iv. The marketing aspect of the study captures only the domestic markets. However, only 30 per cent of the total coffee produced is domestically consumed, while the majority (70 per cent) is exported to foreign countries. The export markets are not covered in the study.

Notes

1 International Coffee Organization. (2019, March). Coffee market report. Retrieved from http://www.ico.org/documents/cy2018-19/cmr-0319-e.pdf.
2 Fairtrade Foundation Coffee farmers. Retrieved from https://www. fairtrade.org .uk/Farmers-and-Workers/Coffee.
3 Fairtrade Foundation. (n.d.). Coffee farmers. Retrieved from https://www. fairt rade.org.uk/Farmers-and-Workers/Coffee.
4 Panhuysen S., and Joost, P. (2018). Coffee Barometer 2018 (p. 36). Retrieved from https://www.hivos.org/assets/2018/06/Coffee-Barometer-2018.pdf
5 Workman D. (2019). Coffee exports by country. Retrieved from http://www. worldstopexports.com/coffee-exports-country.
6 Fairtrade Foundation Coffee farmers. Retrieved from https://www. fairtrade.org .uk/Farmers-and-Workers/Coffee.
7 Sänger, C. (2018). State of the global coffee market. Presented at the 10th Multi-Year Expert Meeting on Commodities and Development, Geneva, Switzerland. Retrieved from https://unctad.org/meetings/en/Presentation/ MYEM2018_Christ oph%20Saenger_25042018.pdf.
8 MMR (2022). Coffee Market: Global Industry Analysis and Forecast (2022-2029). Maximize Market Research Pvt Ltd., Retrieved from Coffee Market: Global Industry Analysis and Forecast (2022-2029) (maximizemarketresearch .com).
9 The Coffee Board estimates yield loss of Indian Arabica up to 40 per cent due to White Stem Borer.
10 One bag of 50 kilos (kg).
11 Currently, the Coffee Board is concerned with the lower productivity of Arabica. Besides, the country fears the extinction of Indian Arabica due to the White Stem Borer. Studies carried out by scientists from the Royal Botanic Gardens, Kew, in 2012 report the extinction of global Arabica coffee by the end of this century.
12 World coffee market is dominated by four major MNCs – Nestle, Kraft, Procter & Gamble and Sara Lee – that control 50 per cent of global markets.

References

Achoth, L., 2005, *Report on surveys on coffee holdings and coffee market chain in India in relation to mould contamination in coffee*. Bangalore: Coffee Board of India; www.fao.org/fileadmin/user_upload/agns/pdf/coffee/Annex-E.2.pdf.

Ady, P., 1968, Supply functions in tropical agriculture. *Bulletin of the Oxford University Institute Economics and Statistics*, 30 (2): 157–188.

Akiyama, T., 2001, *Commodity market reforms: Lessons from two decades*. Regional and Sectoral Studies. World Bank, Washington, DC.

Ambinakudige, S.S., 2006, *Differential impacts of commoditization of agriculture on people's livelihoods and the environment in the Western Ghats of India: An extended environmental entitlement analysis*. Doctoral dissertation, University of Florida; http://purl.flvc.org/fsu/fd/FSU_migr_etd-0190

Arrow, K.J., 1970, The organization of economic activity: issues pertinent to the choice of market versus non-market allocation. In Haveman, R.H. and Margolis, J. (eds) *Public expenditure and public policy*. Markham, Chicago.

Bateman, M.J., 1965, Aggregate and regional supply functions for Ghanaian Cocoa, 1946–1962. *Journal of Farm Economics*, 47(2): 384–401.

Behrman, J.R., 1968, Monopolistic cocoa pricing. *American Journal of Agricultural Economics*, **50**(3): 702–719.

Bhavya, V., 2011, *Value chain analysis for coffee in Karnataka*. M. Sc., thesis. International Master of Science of Rural Development, Ghent University, Belgium.

Business Standard, 2010., *Indian Coffee Planters lose out to Vietnam*. Phillipines.

Chand, Ramesh., 2010, *Estimation of marketing efficiency of horticultural commodities under different supply chains in India*. Research report NCAP, New Delhi, India.

Chatterjee, S., 2005, *Examining the interlinkages between production, consumption and exports of Tea in India*. Proceedings of conference on Empirical Issues in International Trade and Finance, pp. 79–102.

Chattopadhayay, S., and John, P., 2007, *Bitter Beans: The Coffee Crisis and its impact in India*. Partners in Change, New Delhi.

Chengappa, P.G., Rich, K.M., Muniyappa, A., Yadava C.G., Babu, P. and Rich, M. 2014, *Promoting conservation in India by greening coffee. A value chain approach*. NUPI Working Paper 831, Norwegian Institute of International Affairs.

Chethana, A.N., Nagaraj, N., Chengappa P.G. and Gracy C.P., 2010, Geographical Indications for Kodagu coffee – A Socio-economic Feasibility Analysis. *Agricultural Economic Research Review*, **23**(1): 97–103.

Cullis, J. and Jones, P., 1987, *Microeconomics and the public economic: a defense of Leviathan*. Basil Blackwell, Oxford.

David, M. and Clementine, V., 2008, *Coffee Value Chain and Geographical Indications in India: origin, reputation and marketing of Indian coffees*. CIRAD, Agriculture Research for Development.

Devi Uma, K., Panduranga, A., Shareet, S.M. and Srinivasulu, R., 2002, Resource productivity and input use efficiency of coffee in Vishakapatnam District of Andhra Pradesh. *Agricultural Situation in India*, **58**(12): 579–581.

Donnet, Laura, 2007, *The specialty coffee quality rating as a measure of product differentiation and price signal to growers: an entropy analysis of E-auction data*. Annual meeting, July 29–August 1, 2007, Agricultural and Applied Economics Association (AAEA), Portland, Oregon.

Dubois, P., 2001, International cooperation for the development of a sustainable coffee economy. In Chapman, K. and Subhadrabandhu, S. (eds.) *The First Asian regional round-table on sustainable, organic and speciality coffee production, processing and marketing*, 26–28 February 2001, Chiang Mai, Thailand.

Duke, J.A., 1983, *Handbook of energy crops*. Published online. www.hort.purdue .edu/newcrop/duke_energy/Coffea_arabica.html.7p

Dutta, R., 2011, Impact of age and management factors on tea yield and modeling the influence of Leaf Area Index (LAI) on yield variations. *Science Asia,* **37**: 83–87.

Evizal, R., Sugiatno, S., Prasmatiwi, F.E. and Nurmayasari, I., 2016, Shade tree species diversity and coffee productivity in Sumberjaya, West Lampung, Indonesia. *Biodiversitas* **17**: 234–240.

Fernandes, F.L., Picanco, M.C., Campos, S.O., Bastos, C.S., Chediak, M., Guedes R.N.C. and Silva R.S., 2011, Economic injury level for the coffee berry borer (Coleoptera: Curculionidae: Scolytinae) using attractive traps in Brazilian coffee fields. *Journal of Economic Entomology,* **104**: 1909–1917.

Gabre-Madhin, E.Z., 2001, *Market institutions, transaction costs and social capital in the Ethiopian grain market.* Research report 124, International Food Policy Research Institute, Washington, DC.

Guenthner, J.F., Michael, K.C. and Nolte. P., 2001, The economic impact of potato late blight on USA growers. *Potato Research*, **44**(5): 121–125.

Hazarika, K., 2011, Changing market scenario for Indian tea. *International Journal of Trade, Economics and Finance*, **2**(4): 285–287.

Honger, J.O., 2014, *Characterisation of the causal agent of mango anthracnose disease in Ghana.* Doctoral dissertation. University of Ghana, Accra, Ghana.

Indian Coffee Board, 2017, Database on Coffee, March 2017. www.indiacoffee.org

Indian Coffee Board, 2023, Database on Coffee, January 2023. www.indiacoffee.org

Joy, C.V., 2004, *Small coffee growers of Sulthan Bathery, Waynad.* Discussion Paper No. 83, Kerala Research Programme on Local Level Development, Centre for Development Studies, Thirunvanthapuram, http://www.cds.ac.in/krpcds/publication/downloads/83.pdf

Karnik, A., 1997, *Theories of state intervention*, Working paper 96/11, University of Mumbai.

Kirsten, J.F., Dorward, A., Poulton, Colin and Vink, N., 2009, *Institutional economics perspectives on African Agricultural Development.* International Food Policy Research Institute, Washington, DC.

Krishnamurthy, K.S., Kandiannan K., Sibin, C., Chempakam, B. and Ankegowda, S.J., 2011, Trends in climate and productivity and relationship between climatic variables and productivity in black pepper (*Piper nigrum*). *Indian Journal of Agricultural Sciences,* **81**(8): 729–733.

Kumar, P. and Sharma, A., 2006, Perennial crop supply response functions: The case of Indian rubber, tea and coffee. *Indian Journal of Agricultural Economics*, **61**(4): 613–646.

Liebig, Justus Von, 1855, *Principles of agricultural chemistry: With special reference to the Late Researchers Made in England.* Walton and Maberly, London.

McGinnis, M.D., 2011, An introduction to IAD and the language of the Ostrom workshop: A simple guide to a complex framework. *The Policy Studies Journal,* **39**(1): 169–182.

Mesike, C. and Esekhade, T., 2014, Rainfall variability and rubber production in Nigeria. *African Journal of Environmental Science and Technology,* **8**(1): 54–57.

Mohammed, A.B., Ayanlere, A.F. and Ekenta, K.M., 2013, Profitability of coffee in Kabba/Bunu LGA of Kogi State. *African Journal of Agricultural Research*, **8**(23): 2897–2902.

Murphy, T.S., Phiri, N.A., Sreedharan, K., Kutywayo, D. and Chanika, C., 2008, *Integrated stem borer management in smallholder coffee farms in India, Malawi and Zimbabwe.* Final Technical Report. CFC/CABI/ICO/CRS/CB.

Narayana, M.R., 2014, Economic analysis of Leaf Rust Management by chemical controls: Evidence and Implications for household coffee farmers in India. *Indian Journal of Agricultural Economics*, **69**(4): 1–15.

Navendu, V.P., Qureshi Q., Rawat, G.S. and Kushalappa, C.G., 2009, Plant diversity in sacred forest fragments of Western Ghats: A comparative study of four life forms. *Plant Ecology,* **206**: 237–250.

Ngeywo, J.C., Shitandi, A.A., Basweti, E.A., Makone, S.M. and Nyangena, D.N., 2015, Family attachment to coffee farms, a case of coffee farming in Kisii County,

Kenya. *International Journal of Recent Research in Interdisciplinary Sciences*, **2**(1): 1–9.

North, D.C., 1989, Institutions and economic growth: An historical introduction. *World Development*, **17**(9): 1319–1332.

Nsibirwa, R., 2010, *Coffee yield (productivity) and production in Uganda: Is it only a problem of GAP and Diseases?* Forum for Agricultural Risk Management. Available at http://www.agriskmanagementforum.org/sites/agriskmanagementforum.org

Ntimbaa, G.J. and Akyoob, A.M., 2017, Factors influencing choice decision for marketing channels by coffee farmers in Karagwe district, Tanzania. *Global Journal of Biology, Agriculture and Health Sciences*, **6**(2): 1–10.

Oerke, E.C., 2005, Crop losses to pests. *Journal of Agricultural Sciences*, **144**: 31–43.

Pavitha, M.T., 2005, *Post liberalization scenario of Coffee Industry – A case study of Coorg district*. M.Sc (Agri.) Thesis (Unpublished), University of Agricultural Sciences, Dharwad.

Philip, A.P., 2003, *Marketing of spices: A study with special reference to Pepper and Cardamom*. Doctoral dissertation. Mahatma Gandhi University, Kanjirappally, Kerala.

Ponte, S., 2004, *Standards and sustainability in the coffee sector: A global value chain approach*. International Institute for Sustainable Development (IISD) and UNCTAD, Winnipeg, p. 49.

Prabhakaran, P.V., 1997, Quantitative determination of loss in yields of black pepper (*Piper nigrum* L.) in Kannur district (Kerala, India). *Journal of Spices and Aromatic Crops*, **6**(1): 31–36.

Prakash Kammardi, T.N., 2012, *Arecanut economy at the Cross Roads*. Report of the Special Scheme of Cost of Cultivation of Arecanut in Karnataka, Department of Agricultural Economics, UAS, Bengaluru.

Promme, P., Kuwornu J.K.M., Jourdain D., Shivakoti G.P. and Soni, P., 2017, Factors influencing rubber marketing by smallholder farmers in Thailand. *Development in Practice*, **27**(6): 865–879.

Rajkrishna., 1963, Farm Supply Response in India-Pakistan: A Case Study of Punjab Region. *Economic Journal*, **73**(29): 447–487.

Ramaila, M., Mahlangu, S. and Toit, D. 2011. *Agricultural productivity in South Africa: literature review*. Directorate of Economic Services Production Economics Unit, Pretorira. Available at http://www.nda.agric.za/docs/Gen Reports/AgricProductivity.pdf (Accessed 8 September 2014).

Ravindranath, N., 1999, Liberalization of Indian coffee: Does it mean prosperity? *Indian coffee*, **63**(1): 13–15.

Reddy, D.R.B., Shivaprasad, P. and Naidu, R., 2003, Estimation of cost of production of Arabica coffee in Chikmagalur Region, Karnataka. *Journal of Coffee Research*, **31**(2): 106–117.

Rodriguez, B., Andre, R.C. and Burger, K., 2014, *Contractual choice and performance of Mexican coffee traders: The role of market institutions in commodity chains*. Development Economics Group, Wageningen School of Social Sciences, Wageningen University, Hollandseweg, The Netherlands.

Ruben, R., Boselie, D. and Lu, H.L., 2007, Vegetables procurement by Asian supermarkets: A transaction cost approach. *Supply Chain Management*, **12**(1): 60–68.

Rutherford, M.A. and Phiri, N., 2006, *Pests and diseases of coffee in Eastern Africa: A technical and advisory manual.* Technical report. CAB International, Nairobi.

Saini, V., Luhach, V.P., Singh,V.K., Singh, M.M. and Jyoti, V., 2013, Constraints faced by Guava growers in Production and marketing of districts of Haryana state. *Journal of Agriculture and Veterinary Science*, 5(5): 17–20.

Schultz, T.W., 1964, *Transforming traditional agriculture.* Yale University Press, New Haven, CT.

Singh, S., Ladha, J.K., Gupta, R.K., Bhushan, L., Rao, A.N., Sivaprasad, B. and Singh, P.P., 2007, Evaluation of mulching, intercropping with Sesbania and herbicide use for weed management in dry-seeded rice (Oryza sativa L.). *Crop Protection,* **26**: 518–524.

Subervie, J., 2008, The variable response of agricultural supply to world price instability in developing countries. *Journal of Agricultural Economics*, **59**(1): 72–92.

Swamy, D., Chinnappa, B., Ramesh, G. B., Kumar, N.P., 2016, Economic analysis of coffee cultivation in Kodagu district of Karnataka State India. *International Journal of Agricultural and Biosystems Engineering*, **3**: 12.

Titus, A. and Pereira, G.N., 2011, The fine art of irrigation in Robusta coffee Plantations. Available at https://ecofriendlycoffee.org/the-fine-art-of-irrigation-in-robusta-coffee-plantations

Ukers, W.H., 1935, *All about coffee.* 2nd edition. The Tea and Coffee Trade Journal Company, New York.

Umesh, K.B., Vishnuvardhana and Thirumalaraju, G.T., 2005, Business opportunities in cashew industry. *The Integrated Agribusiness Magazine*, 8(1): 14–17.

Velmourougane, K., Panneerselvam, P., Gopinandhan, T.N., Raghuramulu, Y. and Naidu, R., 2004, *Investigations of on-farm processing and drying of coffee in India to develop a GAP and GMP plan for prevention of mould formation.* Proceedings of International Conference on Coffee Science, Bangalore.

Venugopal, M.N., 1995, Viral diseases of cardamom (Eletteria cardamomum Maton). *Journal of Spices and Aromatic Crops*, **8**(2): 185–187.

Vettas, N., 2006, Market control and competitive issues along the commodity value chain. In *Proceedings of the FAO Workshop on Governance, Coordination and Distribution along the Commodity Value Chains, Rome*, 4-5 April 2006, Trade and Markets Division of FAO, Rome, pp. 9–26.

Watheri, M., 2014, *Factors influencing coffee production by small-scale farmers: A case of Tetu constituency.* University of Nairobi, Kenya.

Wickens, M.R. and Greenfield, J.N., 1973, The econometrics of agricultural supply: An application to the World Coffee Market. *The Review of Economics and Statistics*, **55**(4): 433–440.

Williamson, O.E., 1985, *The economic institutions of capitalism: Firms, markets, relational contracting.* Free Press/Collier Macmillan, New York/London.

Willson, K.C., 1999, *Coffee, cocoa and tea.* CABI Publishing, Wallingford, UK; https://search.worldcat.org/title/coffee-cocoa-and-tea/oclc/39360567.

Vanek, J., 1970, *The general theory of labor-managed market economies.* Cornwell University Press, Ithaca, NY.

Yamin, M., 1991, *A reassessment of Hymers's contribution to the theory of the transnational corporation. The Nature of the Transnational Firm.* Routledge, London.

Chapter 2

Database and methodology

The objective of any scientific investigation is to draw some useful conclusions in the study. In order to arrive at the conclusion, it is essential for the investigator to adopt appropriate methods and procedures. Keeping this in view, the chapter is devoted to explaining the database and methodology adopted to fulfil the objectives of the study. These elements are discussed in the following sequence.

2.1 Sources of data

The data on production was obtained through the survey method carried out in Madikeri, Virajpet and Somwarpet tahsils of Kodagu district. The study employs both qualitative (in-depth interviews, extensive field observations and review of documents) and quantitative data collected from several actors in Kodagu for a period between March and August 2016. The nature of the data is cross-sectional, pertaining to the reference period of the 2015–16 agricultural year. A semi-structured, pre-tested questionnaire with both open- and close-ended questions relevant to the objectives was used to collect the required information from 60 Arabica and 60 Robusta growers. During the process, utmost care was taken to explain the objectives of the study to obtain accurate data and to avoid redundancy during the process.

For traders, collectors, hullers, curers, exporters and roasters-cum-retailers, separate checklists were prepared to collect information on buying and selling behaviour, processing and marketing costs, quality aspects and problems incurred by the intermediaries. The interactions with stakeholders (organizations, government, unions and institutions) were unstructured. Several rounds of discussions were held with them to understand their roles and responsibilities pertaining to production and marketing. The secondary data were collected from publications of the International Coffee Organization (ICO) and the Coffee Board of India.

DOI: 10.4324/9781003541721-2

2.2 Sampling design

2.2.1 Selection of the study area

Kodagu district is one of the important coffee-growing districts in Karnataka state. The district alone contributes around 36.88[1] per cent of the country's total coffee production and stands first in both area (102,063 ha) and production (116,550 MT). Additionally, the district produces about 73.65 and 26.35 per cent of the country's Robusta and Arabica production, respectively. Hence, this region was purposively selected for the present study.

2.2.2 Selection of villages and respondents

The survey respondents were identified through a multi-stage stratified random sampling method, based on the latest official records of the Coffee Board and the Codagu Planters' Association (CPA), Madikeri. In the first stage, three Arabica and Robusta liaison zones were selected. In the second stage of sampling, two villages from each of these liaison zones were selected (Figures 2.1 and 2.2). In the third stage, ten growers were randomly selected from each of these villages, classified as small (<2 ha), semi-medium (2–10 ha), medium (10–25 ha) and large (>25 ha) (Table 2.1a). The total sample size was 120 households.

2.2.3 Selection of market functionaries

After interviewing the growers, the market functionaries were identified, and respondents from each category were selected for the purpose of the study, viz. five agents/traders, five coffee depots, five curers, four hullers, five roasters-cum-retailers and five exporters.

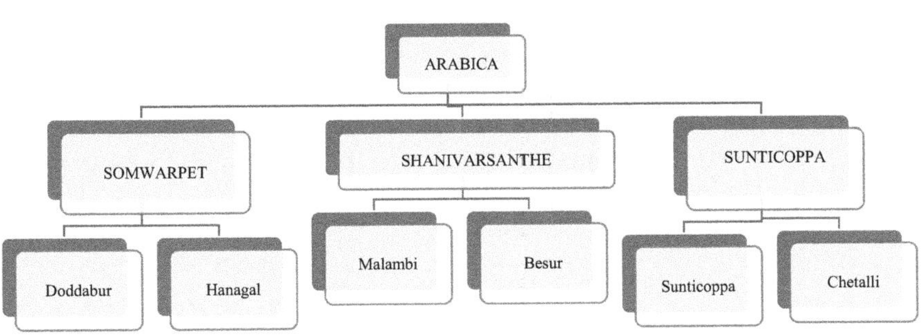

Figure 2.1 Sample villages selected for the study on Arabica and Robusta coffee.

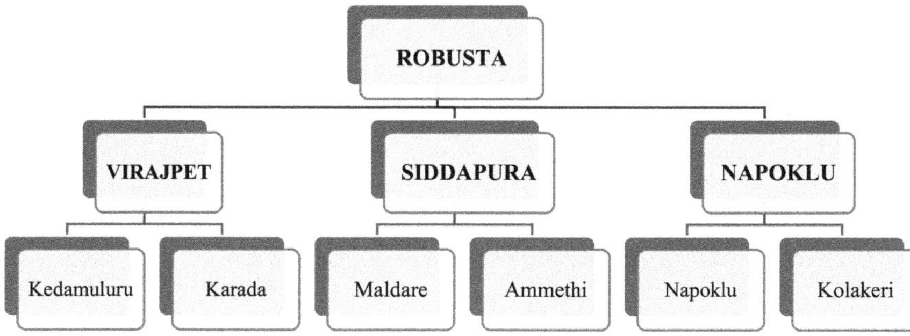

Figure 2.2 Sample villages selected for the study on Robusta coffee.

Source: Own Field Survey.

2.3 The general description of the study area

The Western Ghats of India are exceptionally rich in biodiversity and are recognized as one of the 18 biodiversity hotspots in the world. The globally renowned Western Ghats are also known as 'India's coffee bowl'. These regions account for 97.8 per cent of the total country's production (Indian Coffee Board, 2017). The current study was conducted in Kodagu district, located on the eastern slopes of the Western Ghats. The Kodagu district is situated between 11°50′ and 12°51′ north latitude and 75°25′ and 76°14′ east longitude, covering an area of 410,775 ha (Kodagu District at a Glance, 2016). The district comprises three tahsils, viz. Madikeri, Somwarpet and Virajpet. The forest area comprises 32.76 per cent of the total geographical area of the district. The taluks of Madikeri, Somwarpet and Virajpet are located at altitudes of 1170, 1130 and 909 metres above the mean sea level, respectively. The average rainfall of the district approximately varies between 1500 and 3800 mm. The district comprises 296 villages, with the highest proportion in Somwarpet (45.60 per cent), followed by Virajpet (31.75 per cent) and Madikeri (22.63 per cent). As per the 2011 Census, the total population of Kodagu district was 5.53 lakh, with male and female populations of 2.74 and 2.80 lakhs, respectively. The sex ratio in Kodagu is 1019 females per 1000 males. The overall literacy rate in the district is 82.66 per cent, with male and female literacy rates of 87.19 and 78.14 per cent, respectively (Table 2.2a).

2.4 Details of methodology and data sources followed to fulfil each objective

2.4.1 Economic concepts and analytical tools: the socio-economic and ecological dimensions of coffee production in Kodagu

2.4.1.1 Sources of data

In order to understand the socio-economic and ecological dimensions of coffee production in Kodagu, primary data pertaining to the study were collected from 60 Arabica and 60 Robusta growers from three tahsils of Kodagu.

The primary data on the socio-economic profile of the growers mainly comprised the landholdings of the growers (ha), household size (numbers), average earners (numbers), age of the household members and education status of the growers classified into small (<2 ha), semi-medium (2–10 ha), medium (10–25 ha) and large growers (>25 ha).

The primary data collected on the estate profiles of the surveyed Arabica and Robusta growers were classified based on the elevation of estates (m.a.s.l), annual rainfall (mm), age of coffee trees (years) and land tenures (redeemed and unredeemed land).

In order to study the gross revenue from intercrops, primary data on intercrops (pepper, areca, cardamom, Coorg mandarins and sapotas) were collected from sample estates. This included the area under intercrops (ha), productivity (kg/ha) and gross revenue (₹/ha) from the intercrops.

Further, in order to estimate the total assets of the Arabica and Robusta growers, the primary data on the value of assets (₹ Lakhs) such as farm buildings, labour line, cattle house, storehouse, mud/cow dung yard, cement/tiled yard, weeder, tractors and tillers, sprayers, sprinkler set, pulper and mechanized dryer were collected through a primary field survey. The primary data also included the credit details of the growers, such as the amount of loans borrowed from banks, sources of borrowing and purpose of borrowing.

2.4.1.2 Tabular presentation

The tabular presentation was used to compile the socio-economic profile of the growers, the estate profile of the growers, the intercrops of the estates and the total assets and credit details of the growers. Simple statistical tools such as averages and percentages were used to compare, contrast and interpret the results in the study.

2.4.2 Study of the major factors influencing the productivity of coffee in Kodagu

2.4.2.1 Sources of data

The study used both primary and secondary data. The primary data were collected from the growers for the current study and included productivity of coffee (kg/ha), age of coffee trees (years), elevation of estates (metres), average annual rainfall (mm), rainfall during the critical period (mm), area under coffee (ha), number of shade trees per ha and incidence of White Stem Borer (WSB) in Arabica. The secondary data were used to substantiate the primary data. The secondary data on productivity of coffee (kg/ha), cultivation of coffee under different elevations and ideal rainfall conditions for coffee were collected from the Coffee Board of India. However, the secondary data pertaining to the mean monthly rainfall of the district was collected from the Directorate of Economics and Statistics, Bengaluru. The secondary data for this study were also obtained through newspapers and earlier literature reviews.

2.4.2.2 Tabular presentation

The tabular presentation was adopted to compile data on the productivity (kg/ha) of coffee, age (years) group of coffee trees, elevation (metres) of sample estates, rainfall (mm) at sample estates, optimum, ideal and critical rainfall (mm) conditions for coffee estates, shade trees (number) of estates and area under coffee (ha).

2.4.2.3 Graphical presentation

The graphical representation includes data on the age–productivity of coffee, average annual rainfall, critical rainfall versus productivity, shade tree density and productivity of coffee.

2.4.2.4 Descriptive statistics

The descriptive statistics such as mean, minimum and maximum were computed and presented in tabular format to study the productivity (kg/ha) of coffee, variations in the age (years) of coffee trees, elevation of sample estates, average annual rainfall (mm), critical rainfall (mm), shade trees per hectare (number) and area under coffee (ha).

2.4.2.5 Multiple regression analysis

Multiple regression analysis was used to understand the factors influencing the productivity of coffee. The functional form of the model is given as follows:

$$Y = \alpha_0 + \beta_1 \text{ age} + \beta_2 \text{ age}^2 + \beta_3 \text{ elevation} + \beta_4 \text{ critical rainfall} + \beta_5 \text{ annual rainfall} + \beta_6 \text{ area under coffee} + \beta_7 \text{ shade trees} + \beta_8 \text{ WSB incidence} + \beta_9 \text{ Pruning} + e$$

where Y is the productivity (kg/ha) of coffee
e is the unexplained error term
α_0 is a constant or intercept

2.4.3 Coffee White Stem Borer (Xylotrechus quadripes) and its impact on Arabica variety of coffee

2.4.3.1 Sources of data

In order to understand the influence of Coffee White Stem Borer (*Xylotrechus quadripes*) on Indian Arabica, primary data pertaining to the study were collected from 60 Arabica growers in Kodagu. The primary data included details of the Arabica growers shifting towards Robusta, classification of estates according to the elevation (m.a.s.l), rainfall of estates (mm) and cultural and chemical costs incurred by the growers for control of White Stem Borer.

The secondary data pertaining to the study on the average price of Arabica and productivity of Arabica and Robusta over the years were collected from the Coffee Board of India. Additionally, the secondary data included earlier literature, internet searches and reports on WSB.

2.4.3.2 Tabular presentation

The details of the growers shifting from Arabica to Robusta, elevation and rainfall of estates versus the incidence of White Stem Borer (WSB) and costs incurred for WSB control in Arabica were presented in tabular form.

2.4.3.3 Graphical presentation

The data on the average prices (₹/kg), productivity (kg/ha) of coffee over the years, sources of irrigation, rainfall and incidence of WSB were represented in the graphical format.

2.4.3.4 Likert test

The five-point Likert scales were used in the survey on the opinion of the growers shifting towards Robusta, with respondents indicating whether the respondents strongly agreed, moderately agreed, felt neutral, moderately disagreed or strongly disagreed on shifting from Arabica to Robusta.

2.4.3.5 Multiple regression analysis

The incidence of WSB, in terms of the number of Arabica trees uprooted per hectare, was regressed against the elevation of estates (m.a.s.l), area under coffee (ha), actual rainfall (mm), shade trees per hectare and expenditure on pesticides. The functional form of the model is given below:

$$\ln(\text{number of trees uprooted}) = \alpha_0 + \beta_1 \ln(\text{area under coffee}) + \beta_2 \ln(\text{elevation}) + \beta_3 \ln(\text{actual rainfall}) + \beta_4 \ln(\text{shade trees per ha}) + \beta_5 \ln(\text{expenditure on pesticides}) + \beta_6 d_1 + \beta_7 d_2 + e$$

where Y is the number of Arabica trees uprooted per hectare
α is a constant or intercept
e is the unexplained error term

2.4.4 The economics of Arabica and Robusta coffee cultivation in Kodagu

2.4.4.1 Sources of data

The study on the economics of Arabica and Robusta comprised primary data collected on costs (labour costs, input costs, maintenance costs, processing costs and marketing costs) and revenue from the growers.

2.4.4.2 Definition of terms and concepts used in the study of costs

The procedure to compute the cost concepts is given below.

Human labour – Human labour, including hired and own family labour, required for different operations was expressed in terms of man-equivalent days of eight hours, which were calculated based on the actual paid-out rate.

Fertilizers, manures, insecticides, pesticides and other chemicals – Worked out at the actual prices paid by the cultivators.

Irrigation charges – The actual irrigation charges incurred by the cultivators.

Land revenue, cesses and taxes – The land revenue and other relevant taxes and cesses which were actually paid by the cultivator.

Depreciation on implements and machinery – Farm assets like implements, farm buildings and machinery were evaluated at the prevailing market prices, taking into consideration the collection of assets. Depreciation of these assets for the current year was calculated by using the straight-line method. For this purpose, the present value and the remaining useful life of the assets were considered.

Depreciation = (Purchase value – Junk value) ÷ Average life of asset in years.

Interest on fixed capital – Interest on fixed capital was calculated at the rate of 12 per cent on total working capital at which the banks charge for the short-term loan.

Interest on working capital – Interest on working capital at the rate of 11 per cent per annum, as the crop production is for one full year.

Rental value of land – Rental value of owned land was estimated as one-sixth of the value of gross produce.

Amortized establishment cost – The fixed cost is incurred during the establishment period of coffee. Therefore, it is essential to apportion the fixed costs incurred during the initial three years over the lifespan of the tree in order to obtain the estimates of the total costs incurred during the year. For this reason, we assume that the optimum life period for coffee is 50 years. The discount rate was taken as 10 per cent. The following amortization method was used in order to calculate the annual value of the fixed cost:

$$P = B \frac{i}{1 - (1 + i)^{-n}}$$

where P is the amount of annual payment

B is the initial amount
n is the number of years (life period of plantation)
i is the interest or discount rate (10% in the present case)

Then, the value of the fixed cost was amortized over the life period of the tree to obtain the per annum cost for this one-time expenditure.

2.4.4.3 Multiple regression analysis

$\ln(\text{gross income}) = \alpha_0 + \beta_1 \ln(\text{labour use}) + \beta_2 \ln(\text{fertilizers}) + \beta_3 \ln(\text{irrigation}) + \beta_4 \ln(\text{pesticides}) + \beta_5 \ln(\text{weedicides}) + e$

where Y is the gross income per hectare
α is the constant or intercept
e is the unexplained error term

2.4.5 The market structure and pattern of coffee in Kodagu

2.4.5.1 Sources of data

Apart from the growers, the primary data for the study on market structure and pattern were collected from local agents at the farm gate (5), coffee depots (5), curing works (5), hullers (4), exporters (5) and roasters-cum-retailers (5). The secondary data on farm gate prices, international prices, quantity of shipment of coffee by various ports, value addition and number of licenced curing works were collected from the Coffee Board of India. The data on Codex Alimentarius Maximum Residual Limits (MRL) were collected from the International Coffee Organization (ICO).

2.4.5.2 Tabular presentation

The tabular presentation was used to understand the mode of sales, variations in monthly farm gate prices, certified coffee estates, trends in sustainable production and sales, marketing efficiency and price spread.

2.4.5.3 Descriptive statistics

The descriptive statistics such as mean, minimum and maximum were computed and presented in tabular format to study the variations in farm gate prices (₹/50 kg) and international prices (₹/50 kg). However, the standard deviation and coefficient of variation were used to study the variation in monthly farm gate prices.

2.4.5.4 Marketing efficiency

Acharya's modified marketing efficiency was used to estimate the marketing efficiency of coffee, which is expressed as

$$MME = FP/(MC + MM)$$

where

MME is the modified marketing efficiency
FP is the price received by the farmers/coffee growers
MC is the marketing cost
MM is the marketing margin

2.4.5.5 Price spread

The price spread is the difference between the price paid by the consumer and the price received by the producer, viz. $P_1 - P_2$.

Stationarity test – The existence of the stationarity of time series price data was tested through the augmented Dickey–Fuller (ADF) unit root test.

Cointegration – The Johansen cointegration test has been used to analyze the cointegrating relationship between domestic and international coffee prices. This test was used to understand the long-term relationship existing between the variables.

Granger causality test – The Granger causality test is a statistical hypothesis test for determining whether one time series data is useful in predicting another, exhibiting two types of output, namely unidirectional and bidirectional relationship.

2.4.5.6 Multiple regression analysis

$$\ln(\text{price/kg}) = \alpha + \beta_1 \ln(\text{area under coffee}) + \beta_2 \ln(\text{moisture content}) + \beta_3 \ln(\text{out-turn}) + \beta_4 \ln(\text{total assets}) + \beta_5 \ln(\text{harvesting costs}) + e$$

where Y is price/kg
α is the constant or intercept
e is the unexplained error term

2.4.6 The role of institutions in production, marketing and exports of coffee

Sources of data – The study comprised primary data collected from the growers on the redeemed and unredeemed land tenure systems in Kodagu, productivity (kg/ha) of coffee, number of shade trees per hectare, per cent of silver oaks and shift from Arabica to Robusta under different land tenure systems. The secondary data pertaining to the study were collected from the earlier literature.

Graphical presentation – The bar graph was used to present different types of land tenures among the surveyed growers, productivity, number of shade trees, per cent of silver oaks, shift from Arabica to Robusta under different land tenures and the ratio of producer price to ICO indicator prices. The line graph was used to present Coffee Leaf Rust development seasons and monthly incidence of Coffee Berry Borer.

Note

1 Secondary data pertaining to 2017–18 were collected from the Coffee Board of India.

References

Indian Coffee Board, 2017. Database on Coffee, March 2017. www.indiacoffee.org

Chapter 3

Socio-economic and ecological dimensions of coffee production in Kodagu

The knowledge about the socio-economic background of the coffee growers and the ecological aspects of coffee estates facilitates a better understanding of the problem under investigation. In this study, we collected baseline data on the socio-economic and estate profile of sample growers, which are discussed below. This would also help us examine whether a piece of coffee land could generate sufficient income to sustain the livelihood needs of farm families. Apart from this, the demographic and socio-economic profile of growers plays a pivotal role in determining the major farm practices carried out by the growers, while the ecological dimensions mainly influence the productivity of coffee in the study region.

3.1 Socio-economic profile of the Arabica and Robusta growers

Table 3.1a provides the cross-tabulation of Arabica and Robusta growers, respectively, characterized by the socio-economic characteristics of growers. The results show considerable variations among the characteristics across different-sized growers as follows.

As can be seen from Table 3.1a, the average landholding size with respect to small (<2 ha), semi-medium (2–10 ha), medium (10–25 ha) and large (>25 ha) Arabica growers was about 1.57, 5.19, 14.83 and 90.40 ha, respectively, while in the case of Robusta it was found to be 1.10, 5.60, 18.40 and 45.66 among the small, semi-medium, medium and large growers. On average, the landholding size of Arabica was about 18.15 ha, while that of Robusta was 11.56 ha.

The household size was comparatively larger among the Arabica (5.12) than Robusta growers (4.70). Among the different size holdings, the household size was 6.50, 5.58, 4.07 and 3.62 among small, semi-medium, medium and large growers, respectively. While in the case of Robusta, it was found to be 6.50, 4.64, 4 and 4 among the small, semi-medium, medium and large growers. It can be noted that the household size decreased with an increase in the size of land holdings, unlike other agricultural land holdings where

DOI: 10.4324/9781003541721-3

the relation is generally direct. This was mainly attributed to the spread of Western education in Kodagu and the rise of nuclear families, which was reflected in the prevalence of individual homes and plantation bungalows.

The average number of earners was found to be quite similar in both Arabica (2.38) and Robusta growers (2.36). Likewise, the average number of earners was highest among the small growers (2.87) in both Arabica and Robusta, followed by semi-medium growers in the case of Arabica (2.61) and 2.38 in Robusta. In the case of Arabica, the average number of earners was lowest among the large growers (1.62), whereas in Robusta, the average number of earners was found to be lowest among the medium-sized growers (2.08).

In Robusta, about 50.35 per cent of the total population belonged to the category of earners, while the proportion was relatively less in Arabica (46.58 per cent). On average, the male earners comprised 66.90 per cent of total earners in Robusta, while 33.10 per cent were females. Similarly, in Arabica, 65.73 per cent of total earners were males, whereas 34.27 per cent were females. The per cent of female earners decreased with an increase in the size of land holdings in both Arabica and Robusta varieties.

The age composition of the family members of the selected Arabica and Robusta growers is presented in Table 3.1a. It is evident from the data that there is a concentration of family members in the working age group of 16–60 years. The number of family members in the working-age group (16–60 years) was higher in Robusta (59.22) than in Arabica (56.03). In the case of Arabica, the working age group was highest in medium (67.92), followed by large (65.52), semi-medium (53.20) and small (48.07). However, in Robusta, the working age group was highest in medium and large (66.67), followed by semi-medium (60.12) and small (46.15) growers. Thus, from the analysis, it can be concluded that the majority of the sample respondents come under the working age group.

The percentage of senior members (more than 60 years) was found to be 21.50 per cent in Arabica, while it was 20.21 per cent in Robusta. In both cases, it is evident that there is a concentration of senior members (above 60 years of age) in small growers, viz. 28.85 per cent and 26.92 per cent in Arabica and Robusta, respectively. This was followed by 20.80 per cent (semi-medium) and 20.76 per cent medium growers in Arabica. However, in Robusta, about 26.92 per cent of small, followed by 20.83 per cent of large growers, came under senior citizens.

On average, about 20.57 per cent and 22.47 per cent, respectively, of Robusta and Arabica household members were categorized as children belonging to the age group of less than 16 years. From Table 3.1a, it is evident that about 26 per cent of semi-medium, followed by 23.08 per cent of small, 20.69 per cent of large and 11.32 per cent of medium Arabica growers belonged to this age group. However, in Robusta, the percentage of growers less than 16 years of age was highest in small (26.93 per cent), followed by semi-medium (20.89 per cent), medium (16.67 per cent) and large (12.50 per cent) growers.

The literacy level of the grower is a major factor that determines the adoptability of the method of coffee plantation. Therefore, an understanding of the literacy level of growers becomes essential. The data relating to the levels of literacy of growers reveal that on average about 61.67 per cent of the Arabica growers and 51.67 per cent of Robusta growers were graduates. About 15.00 per cent of Arabica and 23.33 per cent of Robusta growers had education up to the pre-university level, while about 11.67 per cent of both Arabica and Robusta growers were high school educated. There were 5 per cent of Arabica and 1.67 per cent of Robusta growers who studied up to the 7th standard. In both cases, none of the marginal growers pursued post-graduation. Looking at Robusta growers, about 8.33 per cent of medium, 14.71 per cent of semi-medium and 16.67 per cent of large growers were postgraduates. None of the growers were found to be illiterate. From the analysis, it can be concluded that the majority of the sample respondents have completed graduation studies.

Caste-wise distribution of growers reveals that about 48.33 per cent of Arabica growers and 46.67 per cent of Robusta growers belonged to the general category, while about 46.67 per cent of Arabica and 45 per cent of Robusta growers belonged to Other Backward Communities (OBCs). About 5 per cent of Arabica and 8.33 per cent of Robusta growers were minorities. The above table clearly indicates that the majority of growers belonged to the General and OBC categories.

The years of experience is the most important factor determining the adoption of farm practices. Therefore an analysis of the experience level of growers revealed that Robusta growers had more years of experience in coffee cultivation (32.13 years) in comparison to Arabica growers (24.48 years). In Arabica, the medium growers (32.15 years) had the highest years of experience in coffee cultivation, followed by small (26.25 years), semi-medium (22.03 years) and large growers with 19.75 years of experience. Looking into Robusta growers, the small growers had the highest years of experience (36.25 years), followed by large (36.16 years), medium (32.00 years) and semi-medium who had 30.50 years of experience in coffee cultivation.

3.2 Estate profile of growers

The geography of the region has a profound impact on both the production and quality of coffee. In this context, elevation was taken into account. As can be seen in Table 3.2a, the majority of Arabica estates (83.33 per cent) were located at an elevation below 1000 meters above the sea level, while the remaining 16.67 per cent estates located at a higher elevation above 1000 meters above the sea level. About 93.33 per cent of Robusta estates were situated at less than 1000 meters, whereas only about 6.67 per cent of Robusta estates were located above 1000 metres above the mean sea level. Thus, it can be concluded that the majority of sample estates (both Arabica and Robusta) were located at an elevation below 1000 metres.

In terms of rainfall, about 57 per cent of the Arabica estates received rainfall between 1016 and 1778 mm, followed by 43 per cent of estates that received rainfall between 1778 and 2794 mm. About 43 per cent of Robusta estates received rainfall between 2794 and 5080 mm, followed by 35 per cent of the estates between 1016 and 1778 mm. About 22 per cent of the Robusta estates received rainfall between 1778 and 2794 mm.

The age is an important factor that affects coffee production. The percentage of coffee estates comprising different age groups is provided in Table 3.2a. About 26.67 per cent Arabica estates had coffee trees aged between 21–25 years and 26–30 years, while 18.33 per cent of Arabica estates had coffee trees aged 15–20 years, followed by 13.33 per cent of estates with trees aged between 31 and 35 years, and the remaining 3.33 per cent had trees older than 40 years. Looking at Robusta, about 26.66 per cent of estates had trees belonging to the age group of 26–30 years, followed by 20.02 per cent Robusta estates with trees aged between 36 and 40 years, while about 11.66 per cent of Robusta estates comprised trees older than 40 years.

The land tenure system in Kodagu can be broadly classified into redeemed and unredeemed land. In the case of redeemed land, the owner of the land obtains the tree rights. In contrast, in the case of unredeemed land, the government holds the tree rights, though the land actually belongs to the coffee growers (Shrinidhi et al., 2001). About 68.33 per cent of Arabica estates come under the redeemed land tenure system, while the remaining 31.67 per cent fall under the unredeemed land tenure system. About 51.67 per cent of Robusta estates come under unredeemed land tenure, while about 48.33 per cent fall under redeemed land tenure.

3.3 Intercrops in coffee estates

The major intercrops in the study region were black pepper, areca, cardamom, mandarins and sapota, in addition to other minor intercrops (Table 3.3a). The diversification in coffee plantations could help the growers sustain at times when coffee production is not economically viable. Black pepper as an intercrop in coffee plantations earned the maximum gross revenue for the growers. On average, the Arabica growers earned a gross revenue of ₹1.15 lakh/ha from the cultivation of pepper during the reference period 2015–16. The gross revenue earned by the small, semi-medium, medium and large growers worked out to be ₹1.05, ₹1.33, ₹0.93 and ₹0.95 lakhs/ha, respectively. The pepper intercropped with Robusta coffee earned a gross revenue of ₹1.36 lakh/ha. The semi-medium Robusta growers earned the highest revenue from pepper at ₹1.47 lakh/ha, followed by large growers at

₹1.24 lakhs/ha, while medium and small growers earned a gross revenue of ₹1.21 lakhs/ha.

The Arabica growers earned a gross revenue of ₹48,194/ha from areca cultivation, of which the small, semi-medium, medium and large growers' gross revenues were found to be ₹22,462, ₹36,326, ₹66,696 and ₹89850/ha. The areca in Robusta fetched an overall revenue of ₹60,339/ha during the reference period 2015–16. The gross revenue from areca was highest in large growers (₹1.42 lakhs/ha), followed by medium (₹60,149/ha), semi-medium (₹49,576/ha) and small growers (₹44,925/ha).

The cardamom as an intercrop in Arabica earned a gross revenue of ₹29,475 per ha at an overall level, while in Robusta, it worked out to be ₹23,208 per ha. In Arabica estates, gross revenue from cardamom was ₹25,787, ₹30,195 and ₹55371 for semi-medium, medium and large growers, respectively. Looking into Robusta growers, the cardamom fetched the maximum revenue for large growers, viz. ₹46,666 per ha, followed by medium growers, earning a revenue of ₹22,866 per ha, and semi-medium with a gross revenue of ₹18,143 per ha.

The Coorg mandarins fetched an annual gross revenue of ₹7306/ha and ₹12238/ha for Arabica and Robusta growers, respectively, while the revenue worked out to be ₹8599, ₹8284 and ₹8011 for semi-medium, medium and large Arabica growers, respectively. In the case of Robusta, the overall gross revenue from Coorg mandarins was highest in semi-medium growers, earning a gross revenue of ₹14,149/ha, followed by medium (₹14,107/ha) and large growers (₹13,985/ha). For sapotas as intercropping in Arabica plantations, the productivity was highest among the large growers, who earned a gross revenue of ₹25125/ha, followed by medium-sized growers who earned a revenue of ₹18535/ha and semi-medium growers (₹6960/ha). The overall gross revenue from the sales of sapota in Arabica plantations was found to be ₹10396/ha. In the case of Robusta, the gross revenue from intercrop sapotas was highest among small growers (₹19,453/ha), followed by semi-medium (₹18893/ha), large (₹18,250/ha) and medium (₹17,500/ha). The overall gross revenue to the Robusta growers from sales of sapota was found to be ₹18,625/ha.

3.4 Total assets owned by growers

The physical assets owned by the Arabica and Robusta growers are provided in Table 3.4a. The value of total assets was found to be higher among the Robusta growers (₹132.76 lakhs) than Arabica growers (₹127.40 lakhs). Among them, the prominent ones were farm buildings, which constituted 43.09 per cent of value of total assets in Arabica and 46.06 per cent of total assets in Robusta. The coffee drying yard constituted about 20.38 per cent and 17.60 per cent of the value of total assets in Arabica and Robusta, respectively. The coffee pulpers accounted for 12.87 per cent and 5.17 per cent of the value of total assets in

Arabica and Robusta, respectively. The labour line contributed 7.67 per cent and 9.12 per cent of the value of total assets. When taken across the size of holdings, the value of total assets owned by small, semi-medium, medium and large Arabica growers were 14.63, 39.61, 52.63 and 127.40 lakh, respectively. On the other hand, small, semi-medium, medium and large Robusta growers owned assets worth ₹15.10, 45.74, 68.28 and 132.76 lakh, respectively.

3.5 Credit details of growers

Table 3.5a provides the credit details of surveyed Arabica and Robusta growers. It can be noted that all small Arabica growers had bank loans, while 93.55 per cent of semi-medium, 92.31 per cent of medium and 50 per cent of large growers had bank loans. The amount borrowed increased with the size of holdings. About 73.33 per cent of Robusta growers had bank loans. The average amount borrowed among Robusta growers was found to be ₹6.18 lakh. However, none of the small growers borrowed loans from commercial banks while co-operatives formed an important source of credit. The majority of loans (98.33 per cent) were crop loans, while only 3.33 per cent of Robusta growers borrowed loans for the purchase of implements, and about 1.67 per cent of Arabica growers borrowed loans for the purchase of livestock. Overall, it can be concluded that the Arabica growers had more institutional debt than that of Robusta growers.

3.6 Summary

The chapter overviews the socio-economic and ecological dimensions of coffee production in the study region. The socio-demographic characteristics of sample households revealed that apart from the size of land holdings and per cent of earners, there were no marked differences between Arabica and Robusta households, particularly in terms of household size, average number and per cent of earners, age group and education status of growers. The literacy per cent was found to be higher among the surveyed coffee growers in Kodagu. The majority of the growers belonged to general and OBCs, followed by minorities.

The estate profile comprised the elevation of estates, m.a.s.l., average annual rainfall (mm), age of coffee trees (years) and the land tenure system followed in Kodagu. The majority of estates were located below 1000 m.a.s.l., while 57 per cent of Arabica estates received rainfall between 1016 and 1778 mm, and about 43 per cent of Robusta estates received rainfall between 2794 and 5080 mm. The age of Arabica trees varied between 21 and 30 years, whereas about 26 per cent of the Robusta estates had trees in the age group between 26 and 30 years. About 68 per cent of Arabica estates were redeemed. In contrast, about 51 per cent of the Robusta estates were unredeemed.

The black pepper, a major intercrop in the coffee plantations, earned the highest gross revenue for growers, with Robusta growers obtaining the highest revenue (₹1.36 lakh/ha). This was followed by Areca, earning an average gross revenue of ₹60,339 per ha. The gross revenue from cardamom was higher among Arabica growers (₹29,475 per ha) than Robusta growers (₹23,208 per ha). Coorg Mandarins and Sapotas were the other intercrops in the coffee plantations.

The physical assets owned by the growers comprised buildings, labour lines, cattle houses, storehouses, drying yards, tractors and tillers, weeders, sprayers, sprinkler sets, pulpers and mechanized dryers. The overall assets were found to be highest among Robusta growers. Among the assets, the buildings accounted for the highest value (about 42 per cent and 46 per cent of total value of assets in Arabica and Robusta, respectively), followed by cemented/tiled drying yards (20 per cent and 17 per cent of the total value of assets in Arabica and Robusta, respectively). However, about 88.33 per cent and 73.33 per cent of Arabica and Robusta growers, respectively, had bank loans, mainly from cooperatives, followed by commercial banks.

Reference

Shrinidhi, A.S. and Lele, S., 2001, *Forest tenure regimes in the Western Ghats: A compendium*. Institute for Social and Economic Change, Bangalore, **90**: 56.

Chapter 4

Factors influencing productivity of Arabica and Robusta varieties of coffee in Kodagu

4.1 Introduction

There are several factors influencing the productivity of coffee directly or indirectly. An understanding of these factors is crucial for identifying the most limiting or enhancing factors that influence the productivity of coffee. This could also provide policymakers with necessary information about the important factors that influence its productivity, enabling them to eliminate or minimize the limiting factors. At the macrolevel, the study assumes importance in the current context of the ongoing nationwide discussion on how to enhance productivity and improve the production of Indian coffee.

Against this backdrop, the specific objective of the study was to identify the major factors that influence the productivity of coffee. Therefore, we identified seven major factors, viz. (1) age of coffee trees (years); (2) elevation or altitude (metres); (3) average annual rainfall (mm); (4) rainfall during the critical period (mm); (5) area under coffee (ha); (6) number of shade trees per hectare; and (7) incidence of White Stem Borer (WSB) in Arabica (uprooted plants per hectare).

Table 4.1a provides descriptive statistics on the productivity of surveyed estates. The average productivity of Arabica and Robusta varieties of coffee was 803 and 1590 kg/ha, respectively. In other words, there were substantial differences between the productivity of Arabica and Robusta (nearly 1.98 times). The same phenomenon holds true even at the macrolevel (Table 4.2a), which represents that the productivity of Arabica was much lower than that of Robusta. Here, the productivity of Arabica was 1.8–2.3 times lower than that of Robusta in Kodagu and across India in 2020–21. This was mainly attributed to the susceptibility of Arabica to major pests and diseases like White Stem Borer and Coffee Leaf Rust, one of the several reasons responsible for its lower productivity.

Rather than this, within the district itself, there exists a substantial difference in the productivity of coffee. For instance, the productivity of Arabica among the surveyed estates ranged between 600 and 1200 kg/ha, while that of Robusta ranged between 1250 and 1800 kg/ha. Though all the surveyed

DOI: 10.4324/9781003541721-4

estates were situated in the same agro-ecological zone[1] of Karnataka where are the productivity differences occurring? and why are the differences so substantial is a big question. It is obvious that there are certain factors that impact the productivity of coffee, some to a larger extent while some others to a lesser extent. The key question is: what are the major factors that impact the productivity of coffee in Kodagu? This study attempts to find plausible answers to the above questions, which have relevant policy implications for strengthening the coffee industry both at micro and macrolevels.

4.2 Age of coffee trees and productivity of coffee

> The length of life to be expected of many tropical plants, or even their economic or payable life, is subjects on which agricultural science is still remarkably ignorant.
>
> *G.B. Masefield*

Coffee trees are woody perennials grown in the tropics. After planting, it approximately takes about 3–4 years for the plants to blossom and produce fruits, known as gestation period, which also depends on the economic life of coffee trees. Generally, the Arabica shrub attains full bearing and yields economic returns at the age of 8 years, while Robusta takes 10–12 yearsunder good management practices like pruning, irrigation, shade regulation, fertilization and pest and disease management (Titus, 2011). Meanwhile, Duke (1983) notes that the average economic bearing age of coffee trees is 30–40 years, though some plantations bear up to 100 years.

Nevertheless, the age structures of plantations (along with other factors) have a crucial bearing on the productivity levels (Benjamin et al., 1993; Potvin et al.., 2005; Joseph et al., 2010). For instance, the subsequent decline in the production of coffee in several African countries (Angola, Madagascar, Congo, Cote d'Ivoire) under free markets was attributed to ageing coffee trees, while the decline of coffee production in Colombia from 14.1 million bags in 1989–90 to 10.6 million bags in 1990–2000 was majorly due to the existence of old coffee trees (International Coffee Organization, 2014).

Besides, the coffee trees have a well-pronounced age–productivity relation. Therefore, this section estimates the productivity variation of coffee by age since the age determines both the quantity and quality of flowering. An understanding of the age–productivity relation in coffee is of vital importance in the current context of research for several reasons. Firstly, a decline in productivity with the age of coffee implies that the ageing plantations can act as a major constraint in attaining economic sustainability for the grower. Secondly, knowledge of the age-productivity profile is one of the ways to effectively ensure the success of coffee rejuvenation and replanting schemes,[2] as envisaged by the Ministry of Commerce and Industry, Government of

India. Thirdly, the process of modernization of agronomic practices is especially dependent on the age of trees.[3]

Now, in the Indian context, though the age of coffee trees has a strong bearing on their productivity, it is not adequately described in the prior literature. One probable reason could be the non-availability of secondary sources of data on the age of coffee trees, which acts as a constraining factor. For instance, Kumar and Sharma (2006) , in estimating perennial supply response function, point out the non-availability of sufficient age distribution data on perennial crops as one of the major limitations for researchers. Nevertheless, this study attempts to overcome this limitation, as primary data on the age of coffee trees were collected during the field survey.

Table 4.3a presents descriptive statistics on the age of Arabica and Robusta varieties of coffee based on our field survey comprising 120 estates, of which 60 were Arabica and the other 60 were Robusta estates. The mean age of Robusta estates (30.58) was found to be relatively higher than that of Arabica estates (28.02), while the overall mean age of sample estates worked out to be 29.30 years. The minimum age was found to be higher for Arabica (15 years) compared to Robusta estates (10 years). The study was carried out in adult Arabica and Robusta plantations only.

The Robusta estates exhibited a greater maximum age than Arabica; the maximum age for Robusta among surveyed estates was found to be 60 years (Table 4.3a), thereby exceeding the average economic threshold limit (approximately 40 years) of coffee trees. On the other hand, the maximum age of Arabica trees was found to be 46 years. The maximum age indicates the presence of old age/senile coffee estates in 11.66 per cent of Robusta and 3.33 per cent of Arabica estates; thus 'age' is identified as one of the major 'limiting factors' in attaining higher productivity in these estates.

Table 4.4a details the productivity of coffee sorted according to different age classes. In the case of Arabica, productivity increased with age of the coffee trees. The highest productivity was observed between 21 and 25 years of age, viz. 946 kg/ha. After reaching a maximum, the productivity gradually declined thereafter. The least productivity was observed above 40 years of age. Likewise, in case of Robusta, productivity increased with age. The maximum productivity of 1718 kg/ha was observed between the age group of 26–35 years, which declined thereafter.

In our study, we observe a strong inverse U-shaped relation between the age of Arabica trees and their corresponding productivity (Figure 4.1). In other words, the results imply a parabolic relationship, wherein productivity steadily increased from 15 years (700 kg/ha)with high productivity between 20 and 26 years of age (921 kg/ha), but steadily declines after 26 years up to 34 years (823 kg/ha). The productivity was almost stable from 34 to 37 years but gradually declined later up to 40 years (Figure 4.1). Thus, the risk of declining productivity was associated with the ageing of coffee trees; viz. least productivity was observed in Arabica above 40 years of age.

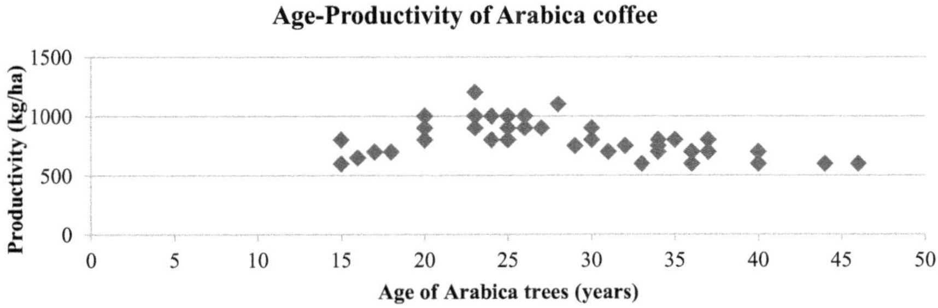

Figure 4.1 The inverted-U-shaped relationship between age (years) and productivity (kg/ha) of Arabica.

Source: Field survey.

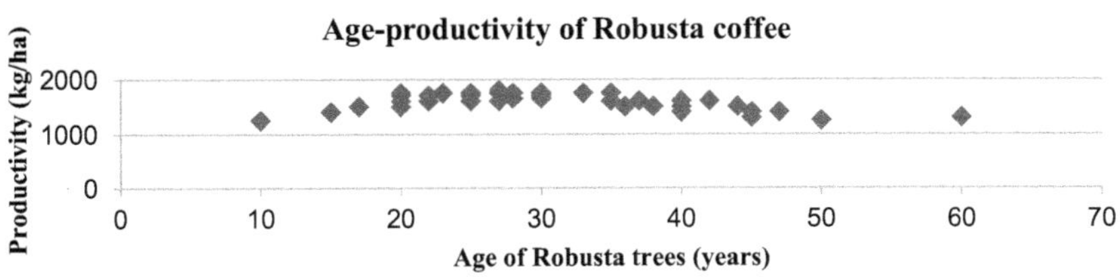

Figure 4.2 The inverted-U-shaped relationship between age (years) and productivity (kg/ha) of Robusta.

Source: Field survey.

The age–productivity relation in the Robusta variety of coffee is provided in Figure 4.2. In the study region, it is found that the productivity of Robusta starts increasing from 10 years of age and attains a maximum between 20 and 30 years of age. A gradual decline in productivity was observed after 35 years of age. This declined productivity was almost stable from 35 to 40 years and subsequently till 50 years, thereby indicating a parabolic relation between age and productivity. It was also interesting to note that the inverse U-shaped relationship was more prominent in Robusta than Arabica.

4.3 Elevation and productivity of coffee

The Kodagu district falls under the hilly agro-climatic zone of Karnataka, known as the Malnad region, characterized by undulating elevations and topographies, which extend between the Western Ghats. The average elevation of Madikeri (the capital of Kodagu) is about 1166 metres above the mean sea level (m.a.s.l.). The highest elevation is Tandiandamol peak, situated at an elevation of 1908 metres above the mean sea level (Kodagu District

Census Handbook, 2011). Nevertheless, coffee is cultivated in the district at elevations ranging between 750 and 1100 m.a.s.l. (Coffee Board of India, 2017).

Generally, the higher elevations are associated with higher qualities (Figure 4.3). The higher elevations are characterized by lower air temperatures, which slows down the coffee ripening process and have a significant impact on the biochemical composition of the beans (chlorogenic and fat concentration). Besides, the slower maturation process of the beans at a higher elevation provides time for complex sugar formation and a complete bean-filling process (Vaast et al., 2006), resulting in denser beans and more intense flavour than the ones cultivated at lower altitudes. Thus, the slower maturation process of coffee beans at a higher elevation plays a pivotal role in determining the quality of coffee (Silva et al., 2005).

From the earlier discussions, it was evident that elevation played a vital role in determining the quality of beans. Likewise, several studies (Guyot et al., 1996; Avelino et al., 2007; Sridevi et al., 2013) have mostly focused on the influence of elevation on the quality of beans, and studies (Hajian et al., 2014; Tolessa, 2016) focused on the influence of elevation on species diversity. However, not many studies have been conducted on the influence of elevation on the productivity of coffee, even though elevation is a powerful factor that influences productivity. Therefore, the impact of elevation on

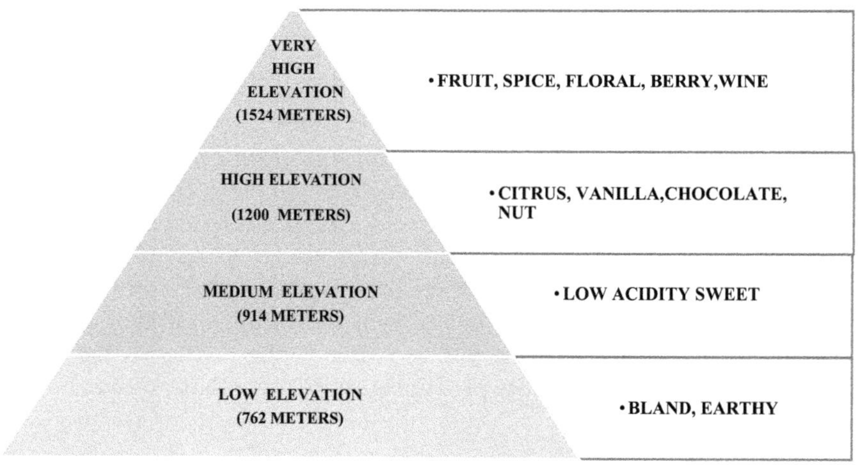

Figure 4.3 Coffee elevation range and quality chart. The magnificent quality coffee reputed from Ethiopia, Kenya, Colombia, Guatemala and Papua New Guinea is primarily attributed to very high altitudes of 1524 m.a.s.l. (5000 ft). Costa Rica, Java, Sumatra, Nicaragua and Mexico grow coffee at an altitude of 1200 m.a.s.l. Coffees at an altitude above 1200 metres are classified as 'Strictly High Grown coffee', commanding high prices. Indian coffee comes under medium elevation (914 metres).

Source: http://www.scribblerscoffee.com/flavor_effect_of_altitude.

productivity is apparent, yet it still remains unexplored, unclear and a complicated one, as both the varieties tend to thrive best at different altitudes. Now, "Does varying elevations significantly influence the productivity of coffee?" If yes, to what extent is the key question. Our study attempts to provide plausible answers to the above questions. Besides, the study would aid policymakers in providing 'location specific policies' even at a macrolevel, as coffee is cultivated at different altitudes across the country, as shown in Table 4.5a.

The Kodagu region on the eastern slopes of the Western Ghats has natural conditions to produce medium-altitude, good-quality coffee. The average elevation among the surveyed Arabica estates (947 m.a.s.l.) was found to be higher than Robusta (924 m.a.s.l.). By and large, the surveyed estates were located in medium elevated (936 m.a.s.l.) mountain ranges of the Western Ghats, while the Robusta estates were located at elevations between 704 and 1072 m.a.s.l. and Arabica estates between 853 and 1026 m.a.s.l. (Table 4.6a).

4.4 Average annual rainfall and productivity of coffee

Coffee is a tropical, climate-sensitive crop, wherein both the varieties require a specific set of climatic conditions to thrive and yield a better harvest. Other than the elevation, coffee trees are also highly sensitive to the rainfall conditions of the region (DaMatta, 2006), due to its effect on temperature, sunshine and soil saturation. The untimely rains are the most important limiting factor for the productivity of coffee, making it vulnerable to climate change and climate variability (Balaji, 2010).

Nevertheless, there have been inherent complexities and uncertainties regarding the influence of rainfall on the productivity of coffee. For instance, studies conducted by Msuya (2013) indicated a weak correlation between rainfall and production of coffee and that coffee production is not influenced by rainfall. In contrast, Jassogne (2013) revealed a declining productivity associated with lower rainfall. However, Ngabitsinze (2011), Thorpe et al. (2012) and Craparo (2012) documented lower productivity associated with heavier rains. Thus, the earlier literature documented contrasting results on the influence of rainfall on the productivity of coffee. Most of these studies may exert generalized effects on the broader responses of coffee without classifying them into Arabica or Robusta, even though each variety requires different rainfall ranges. Therefore, this study attempts to understand the influence of rainfall per annum on the productivity of Arabica and Robusta varieties of coffee in the short run. In order to do so, we have obtained rainfall data from individual estates through a primary survey for the year 2015–16 and tried to explore its impact on the productivity of both Arabica and Robusta varieties of coffee.

The Kodagu district receives a major portion of the rains from the Southwest monsoons (June–September) and the Northeast monsoons

(October–December). The district comes under the hilly agro-climatic zone of Karnataka, characterized by heavy rainfall and the third highest rainfall-receiving district in the state during 2015–16, which received about 2431 mm annual rainfall, only next to Udupi (3771 mm) and Dakshina Kannada (3147 mm) (Figure 4.4). However, the normal rainfall in the district is 2693 mm, thereby the rainfall during 2015–16 was deficient by 9.72 per cent (Directorate of Economics and Statistics, 2015).

As discussed earlier, the district received an average annual rainfall of about 2431 mm during 2015–16, with about 39 mm during the cold weather period (January–February), followed by the hot weather period (March–May) at 1123 mm, Southwest monsoons (June–September) at about 1175 mm and North-east monsoon (October–December) at about 94 mm (Figure 4.5). Interestingly, during 2015–16, the district received only 47.58 per cent rainfall from the South-west monsoons (normally it receives 75 per cent rains), followed by 46.84 per cent rains during the hot weather period.[4] The growers' opinion about the deficit in rainfall during 2015–16 was attributed to the El-Nino effect.

The mean annual rainfall (mm) in the surveyed estates varied between 1229 mm and 3534 mm during 2015–16 (Table 4.7a). The Robusta growing zones significantly received heavier rainfall, viz. 1.38 times more than

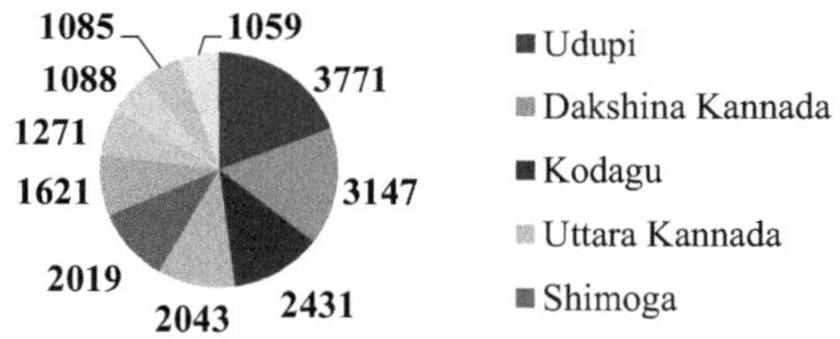

Figure 4.4 Top ten highest rainfall (mm) districts of Karnataka (2015–16).
Source: Directorate of Economics and Statistics, 2015.

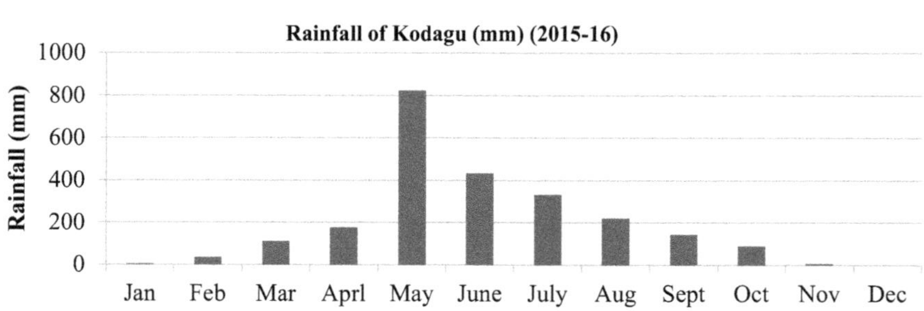

Figure 4.5 The mean monthly rainfall (mm) in Kodagu during 2015–16.
Source: Directorate of Economics and Statistics, 2015.

the Arabica zones, as the majority of the Robusta estates were located in the western and central parts of Kodagu (Napoklu, Madikeri and Virajpet zones of Kodagu), characterized by heavy rainfall as the mountains in the Western Ghats intercept the rain-bearing Westerly monsoon wind. The dense forest in the Western part of Kodagu also contributes to the high orographic precipitation, thereby resulting in heavy rains, viz. about 6250 mm along the Western border. On the other hand, the eastern part of Kodagu lies in the rain shadow belt, thereby receiving far less rainfall, viz. averaging about 1120 mm.

The coffee growers in the region opined that, though the quantum of rainfall has not changed much over the years, the rainfall is not well distributed throughout the year. However, the conditions were worse in the eastern and northeastern zones of Kodagu, especially in the prime Arabica growing belt of Somwarpet taluk (Somwarpet, Shanivarsanthe and Sunticoppa). The taluk produces approximately 12 metric tonnes of Arabica annually. The sporadic rainfalls have made the region dry, thereby making the Arabica crop vulnerable to Coffee White Stem Borer (WSB) attacks. Besides, the growers in the region also face an acute shortage of water since the wells and bore-wells have already gone dry.

Table 4.8a classifies the productivity of Arabica and Robusta coffee in accordance with the rainfall received during the study period. As discussed earlier, the rainfall in Kodagu during 2015–16 was deficit by 9.72 per cent. These low rainfall conditions and resulting dry spells have not only impacted the productivity of coffee but also influenced the production practices followed in coffee cultivation. For instance, in the case of Arabica, the growers observed that the fruits usually ripen in the month of November; however, during our survey, it was found that the fruits started ripening in the month of September itself, thereby resulting in the premature ripening of berries. Due to this, the harvest season also gets preponed due to the prevailing heat conditions existing in the Arabica regions. The same holds true in the case of Robusta too.

It is apparent from Figure 4.6 that a linear relationship existed between rainfall (mm) per annum and the productivity (kg/ha) of Arabica. The average productivity of 800 kg/ha was observed at rainfall ranging between 1400 mm and 2300 mm, with the maximum productivity being almost 1000 kg/ha, at rainfall ranging between 1651 mm and 2286 mm. This rainfall range is in close proximity to the optimum rainfall (1600–2500 mm) conditions for Arabica as provided by the Coffee Board. However, the earlier literature defines different optimum rainfall conditions for Arabica as shown in Table 4.9a, though Arabica is cultivated at different rainfall ranges between 1000 and 1200 mm in Araku Valley (Andhra Pradesh), Anamalais (2500–3000 mm), Bababudangiris (1750–2200 mm) and Biligiris at 1100–1200 mm of annual rainfall.

The variation in the productivity (kg/ha) of Robusta along with rainfall (mm) in the study region is shown in Figure 4.7. The optimum productivity of

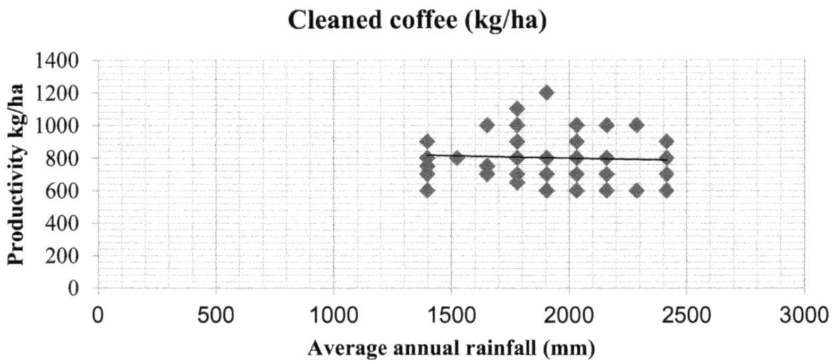

Figure 4.6 Average annual rainfall (mm) and productivity (kg/ha) of Arabica (2015–16).

Source: Field survey.

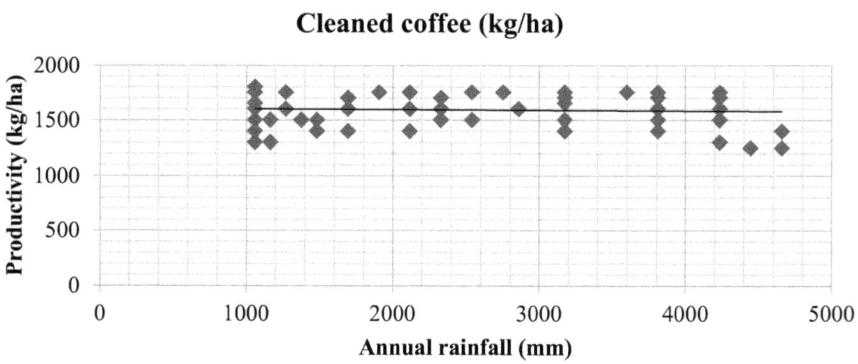

Figure 4.7 Annual rainfall (mm) and productivity (kg/ha) of Robusta coffee.

Source: Field survey.

1600 kg/ha was found between 1058 and 2116 mm annual rainfall, declining thereafter. In contrast to this, Coste (1992) notes the optimum rainfall between 2000 and 3000 mm. While the Coffee Board noted the ideal rainfall conditions for Robusta as being 1000–2000 mm. Interestingly, the least productivity of 1250 kg/ha was observed in estates receiving heavy annual rainfall.

4.5 Rainfall during critical period[5] and productivity of coffee

In the earlier section, we have discussed the impact of annual rainfall on the productivity of coffee in the short run. Further, the timing and quantity of rainfall are also highly crucial for attaining higher productivity, particularly in the Robusta variety of coffee, as it is highly sensitive to water requirement conditions. However, the Arabica, being a drought tolerant crop, does not require irrigation during this period.

In the case of coffee, a certain amount of rainfall is required during the critical period (blossom and backing showers). Blossom showers control the flowering, berry expansion and dry matter formation[6] (Mitchell, 1985), followed by backing showers. In the case of untimely occurrence of rains during this period, it should be supplemented by an artificial source of sprinkler irrigation; otherwise, it would result in the drying up of buds (pinking of buds), thereby proving to be detrimental to the following year's productivity.

By and large, the ideal set of conditions for coffee are characterized by uniform rains throughout the year, followed by 1–2 months of dry spells during the harvest period, viz. December and January. These dry spells are highly crucial as they trigger the hormone[7] in the coffee that induces coffee blossoms at the start of subsequent rains (usually February–March). The quantum and timing of critical rains for Arabica and Robusta are shown in Table 4.10a. With this backdrop, in the following section, we attempt to understand to what extent the rains during this critical period impact the productivity of coffee, as very few studies have been attempted to understand the impact of 'critical rains' on productivity, despite the global importance of 'critical rainfall', a triggering factor for coffee production.

During 2015–16, the majority of the Arabica and Robusta estates received timely and adequate blossom and backing showers. The Robusta estates located at Siddapur and Madikeri and the Arabica estates at Sunticoppa and Chettali zones particularly benefited from these showers. Unlike 2015–16, if the rains failed during this period, the growers had to rely on sprinkler irrigation. However, about 16.67 per cent of the growers pointed out that they were unable to provide sprinkler irrigation owing to financial inefficiency to carry out costly sprinkler operations who in turn relied completely on the blossom and backing showers. On the other hand, about 21.66 per cent of the growers who own sprinkler machinery opined that they do not wait for blossom showers but rather continue with sprinkler itself, so as to ensure higher productivity the following year.

The critical period rain (blossom and backing showers) requirement was more pronounced in the case of Robusta. As observed from Figure 4.8, a considerable difference between productivity in Robusta estates below and above 130 mm critical rainfall was observed. This was more pronounced among the medium growers. Interestingly, the larger estates observed higher productivity below 130 mm. The probable reason could be that the large Robusta growers usually do not wait for the blossom showers and carry out sprinkler irrigation, irrespective of blossom showers to induce good blossoms. However, the truth could be that the overhead sprinkling of larger

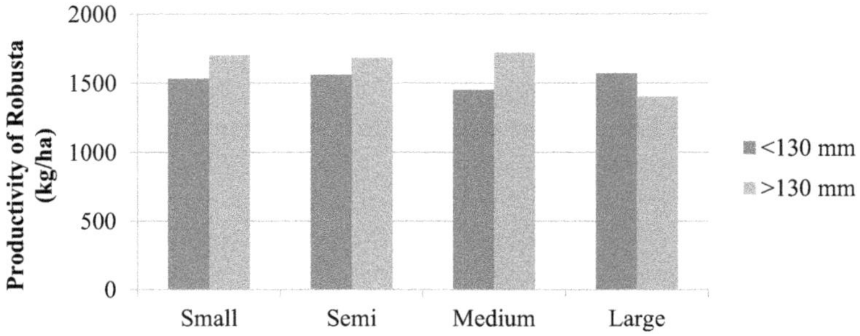

Figure 4.8 Productivity of Robusta vs rainfall during the critical period in accordance with the size of land holdings.

Source: Field survey.

estates could have brought about a sudden drop in temperature and that the critical changes in the microclimate could have induced good blossoms in the initial days. This was the case for larger estates with ≤130 mm rainfalls.

However, the productivity was found to be lower in the larger estates that received >130 mm of critical rainfall. With the reception of natural showers (backing showers) in the latter days, the estates unusually receive higher than the optimum amount of moisture. This could result in a heavy crop load and dropping of the newly set fruits, thereby adversely impacting Robusta estates that received more than 130 mm of critical rains. In other words, the coffee bushes require only a certain amount of water while providing water in excess of the optimum requirement could lead to physiological damage in the flowering and fruit setting.

Though the sprinkler irrigation to Robusta estates could provide substantial economic benefits to the growers, this could also have an adverse ecological influence on the micro-environmental set-up. For instance, during the earlier years, the conventional high shade coffee production was a combination of natural blossom showers (and backing showers) and indigenous mixed shade trees that promised good returns. However, the natural rains, when being supplemented by an artificial sprinkler irrigation, have resulted in a sharp decline in the surrounding temperature of the region. The thick shade canopies from indigenous trees in coffee estates gradually lost their significance and are being replaced by silver oaks.

The increasing trend in the productivity of Robusta with critical rainfall (blossom and backing showers) was recorded at Ambalavayal (Figure 4.9). It was observed that, during 1983, there were no blossom rains either in February or March, which resulted in poor flower setting, while the following year, a higher shower during the critical period resulted in higher fruit setting that was reflected in terms of higher productivity.

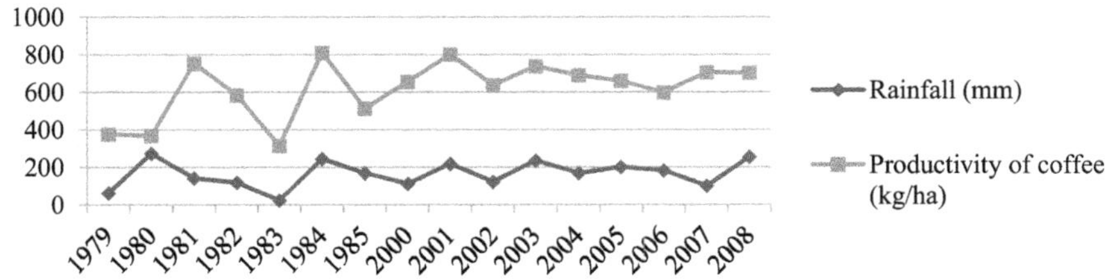

Figure 4.9 Critical period rainfall and productivity of coffee at Ambalavayal, Waynad and Kerala.

Source: Gopakumar, 2011.

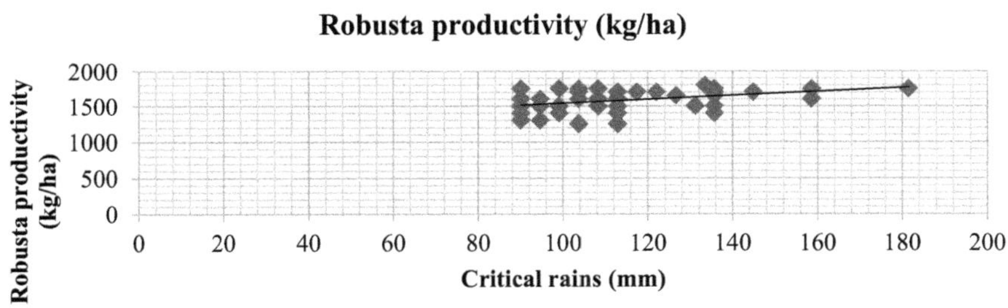

Figure 4.10 Linear relation between critical rains and productivity of Robusta.

Source: Field survey.

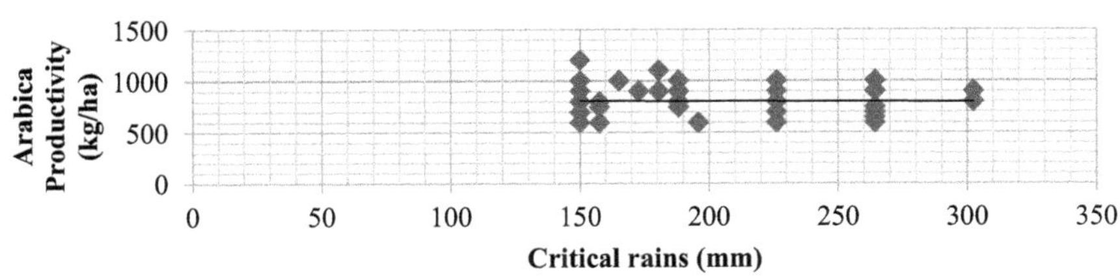

Figure 4.11 Linear relation between critical rains and productivity of Arabica.

Source: Field survey.

As seen from Table 4.11a, the productivity of Arabica reduced with the critical rains, reached a minimum (755 kg/ha) with the rainfall range between 251 and 300 mm, and then with the increase in critical rains above 300 mm, the productivity of Arabica increased to 820 kg/ha. In the case of Robusta, productivity increased with the increase in critical rains (Figure 4.11).

The rainfall deficit during the critical period could be considered a limiting factor on productivity. The water deficits during this period affect the net photosynthesis process and reduce stomatal conductance, enzymatic activities and also the metabolism of carbon (Covre et al., 2016), thereby reducing the number of flower buds, the number of plagiotropic branches per plant, nodes per branch and fruits per node, with overall reduction in production.

On the other hand, Figure 4.11 indicates the linear relation between critical period rains (mm) and productivity (kg/ha) of Arabica. The optimum productivity of 800 kg/ha was observed between 160 and 300 mm of critical rainfall. The productivity variation was observed in estates receiving an equal quantum of critical rains, which was mainly attributed to pest and disease management, number of shade trees, elevation, annual rainfall and other agronomic practices adopted by the growers.

4.6 Shade trees and productivity of coffee

Indian coffee is a shade-grown coffee, traditionally cultivated under a thick canopy of shade. Over the past few years, however, there has been a conversion from the traditional shaded system to an un-shaded system in response to external market dynamics. Coffee growers, aiming to increase productivity, have started intensive cultivation of coffee by reducing the number of shade trees. The growers are shifting from shade to partially sun-grown coffee. The native trees are being replaced by exotic silver oaks, as the growers perceive that too many shade trees reduce the productivity of coffee. Owing to the removal of shade trees in coffee estates, Garcia et al. (2010) document that from 1970 about 33 per cent of the forest cover has been lost in Kodagu.

Now, coming to the type of shade systems existing in coffee estates, Figure 4.12 shows different shade systems in coffee plantations. The 'rustic forest systems' are the ones where only the lower strata of indigenous trees are removed for cultivating coffee while the original native trees are maintained. Traditionally, coffee was grown in Kodagu under rustic shade. Similarly, in the traditional polyculture system, coffee is cultivated under the original forest canopy at lower strata with diversification of beneficial shrubs for medicinal,

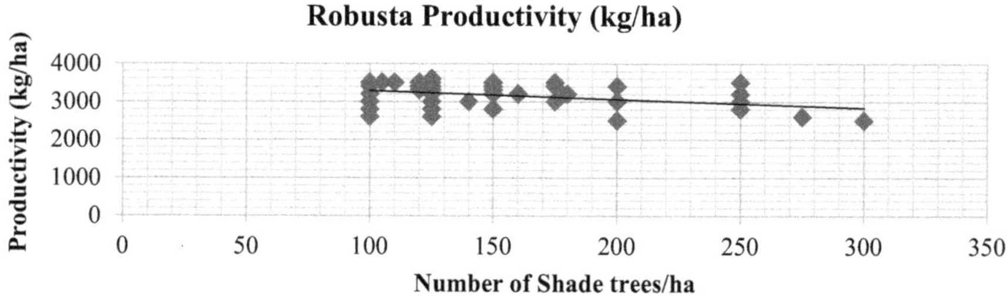

Figure 4.12 Linear relation between shade tree density and productivity of Robusta.

Source: Field survey.

food and fuel purposes, which stood as a risk mitigation strategy in times of depressed coffee prices. In the case of commercial polyculture, a certain per cent of canopy trees is removed, thereby comprising only two strata, viz. the top layer comprising the forest canopy and the bottom strata comprising coffee plantations. The shade monoculture comprises a well pruned, single canopy of species (usually leguminous plants) solely for commercial purposes. On the other hand, the unshaded monoculture or full-sun exposes coffee shrubs to direct sunlight and higher chemical inputs that could have greater environmental costs (Moguel and Toledo, 1999; Jezeer et al., 2017).

There are research studies (Muschler, 2001; Santos et al., 2012; Somporn et al., 2012; Evizal et al., 2016) which have documented a positive relation between shade and the productivity of coffee. On the other hand, Morais et al. (2006) and Vaast et al. (2006) have noted an inverse relation between shade and productivity. Alvarado et al. (2002) and Ricci et al. (2011) show no changes in the productivity of coffee due to shade. The earlier literature on the relation between shade and productivity has shown contrasting results, probably due to differences in biophysical environments, plant materials and agronomic practices followed by the growers. Although coffee agroforestry systems have been studied in isolation, this research would look from an economic perspective at the relation between shade and the productivity of coffee. The key question being to what extent the shade trees affect the productivity of different varieties of coffee in Kodagu.

The descriptive statistics pertaining to the number of shade trees per hectare is provided in Table 4.12a. The average shade tree density stood at 192 trees/ha, while the Arabica estates maintained a higher shade density (237 shade trees/ha) compared to Robusta (147 shade trees/ha). In general, the number of shade trees per hectare varied between 100 and 375 trees per ha (Table 4.12a). In India, the shade canopy, being dense and diverse, supports rich biodiversity (Bandeira et al., 2005).

Nevertheless, Ambinakudige et al. (2009) note that the land tenure system existing in Kodagu is one of the major reasons for conserving native trees, viz. redeemed and unredeemed land tenures in Kodagu. In the redeemed land, the owner of the land holds the tree rights, whereas in unredeemed land, the government holds tree rights, even though the land is owned by the grower (Shrinidhi et al., 2001) and permission is mandatory to sell native trees, thereby restricting the growers from cutting the trees on their own estates. This tree-cutting restriction could be one of the reasons for conserving the trees in Kodagu.

The productivity of coffee classified as per the number of shade trees per hectare is provided in Table 4.13a. The productivity of Arabica was found to be lower (792 kg/ha) with <200 shade trees per ha. The productivity gradually increased with the number of shade trees. The maximum productivity in Arabica was observed with the number of shade trees between 300 and 400 shade trees per ha. However, productivity decreased (783 kg/ha) with >400

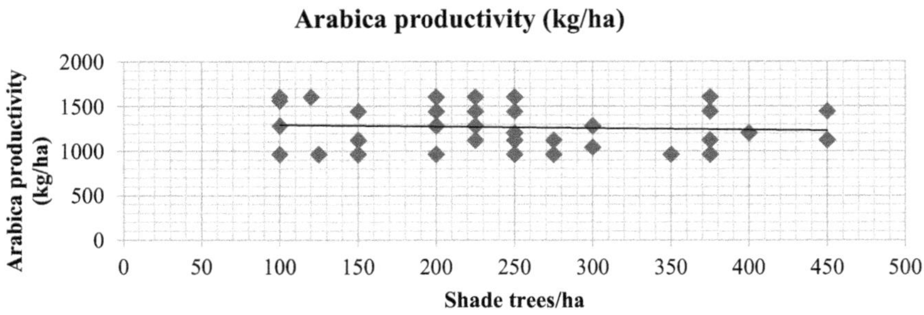

Figure 4.13 Linear relation between shade tree density and productivity (kg/ha) of Arabica.

Source: Field survey.

shade trees per ha. However, in case of Robusta, the maximum productivity of 1630 kg/ha was found with 150 to 200 shade trees per ha. The productivity of Robusta was found to be lower with >250 shade trees per ha.

There exists a negative relation between shade and the productivity of Robusta (Figure 4.12). Jaramillo et al. (2013) note that the number of nodes and productivity of coffee decrease with shade. The limitation of photosynthetically active radiation (PAR) causes a lesser number of plagiotropic[8] branches. This could lower the productivity of coffee.

The experiments conducted by Adugna et al. (2011) reveal that productivity was lower in shade-grown coffee. However, the shade trees protect coffee trees from adverse environmental stress, higher soil temperature and lower relative humidity, besides activating the biochemical and physiological behaviour of coffee trees to facilitate better photosynthetic activities. Though our study reveals an inverse relation between shade and productivity, it is found to be less pronounced in Arabica than in the Robusta variety of coffee (Figures 4.12 and 4.13); as a matter of fact, Arabica requires more shade than Robusta (Chengappa et al., 2014a).

4.7 Area under coffee and productivity of coffee

An inverse relationship between farm size and productivity in agriculture is well established owing to the intensive cultivation of land by smallholders. Now the key question is, "Does an inverse relation between farm size and productivity exist in coffee" as well. Therefore, in this section, we attempt to establish the relationship between farm size (area under coffee) and productivity in both the varieties of coffee.

In order to illustrate the relation between farm size (area under coffee (ha)) and productivity of coffee (kg/ha), we divided the area under coffee into two equal groups, viz. Arabica and Robusta estates below and above 10 hectares. On average, the sample households owned 14.86 ha of land, ranging between 0.90 and 115.20 hectares. In general, the area under coffee was higher among the Arabica growers (18.16 ha), with the minimum and maximum areas under Arabica varying between 1.20 and 158 ha, respectively. The areas under Robusta among the surveyed estates were found to be 11.57 ha, with the minimum and maximum values ranging between 0.60 and 72 ha, respectively (Table 4.14a).

It can be observed from Table 4.15a that maximum productivity in Arabica (833 kg/ha) was observed between 26 and 50 ha land holdings, while the least productivity was observed below 2 ha land. In the case of Robusta, productivity increased with the size of holdings, reached a maximum between 21 and 50 ha land holdings and then declined with the increase in holdings.

4.8 Factors influencing productivity of Arabica: regression analysis

The functional form of the model is given as

$$Y = \alpha_0 + \beta_1 \text{ age} + \beta_2 \text{ age}^2 + \beta_3 \text{ elevation} + \beta_4 \text{ critical rains} + \beta_5 \text{ annual rainfall} + \beta_6 \text{ area under coffee} + \beta_7 \text{ shade trees} + \beta_8 \text{ WSB incidence} + \beta_9 \text{ Pruning} + e \text{------------------------------} \quad (4.1)$$

where Y is the productivity (kg/ha) of Arabica
e is the unexplained error term
α_0 is a constant or intercept

In the above equation, the square term[9] was added to the independent variable, viz. age. By doing so, the quadratic effect of regression is known under the assumption that the effect of age on productivity changes (positive or negative) after a certain point. As seen from Table 4.16a, the coefficient with respect to age was found to be positive (29.07), while the age square was negative (–0.59) and was found significant at 1 per cent. This means that the productivity of Arabica initially increased with plant age, but after reaching the peak, it starts declining with age. For instance, the first three years of the coffee life cycle is an unproductive period, while the harvest of coffee begins from the fourth year onwards, though lower productivity was observed during this period. Generally, Arabica productivity becomes high and stable approximately at the age between 21 and 26 years, after which a gradual decline in productivity is observed, as shown by the inverse 'U'-shaped relation between age and productivity (Figure 4.2).

The elevation of Arabica estates had a positive influence on their productivity. As seen from Table 4.16a, the coefficient associated with elevation was found to be positive, viz. 0.36, and significant at the 1 per cent level of significance. In other words, for every 1 per cent increase in the elevation of coffee estates, the productivity of Arabica increases by 0.36 per cent.

The regression results indicate that the area under Arabica turned out to be positive (0.65) and significant at the 5 per cent level of significance. Similarly, the costs of pruning Arabica estates also had a positive influence on productivity and were significant at the 1 per cent. Likewise, the number of shade trees in coffee estates had a positive influence on the productivity of Arabica, i.e. the coefficient associated with the shade trees was found to be 0.24. In contrast to this, the incidence of WSB (uprooting of Arabica trees/ha) was found to be negative and significant at the 1 per cent. On the other hand, the coefficient associated with critical rains in Arabica and average annual rainfall turned out to be insignificant in the model. Overall, the model was significant at the 1 per cent and the associated R^2 was found to be 0.80.

4.9 Factors influencing productivity of Robusta: regression analysis

The functional form of the model is given as follows:

$$Y = \alpha_0 + \beta_1 \text{ age} + \beta_2 \text{ age}^2 + \beta_3 \text{ elevation} + \beta_4 \text{ area} + \beta_5 \text{ annual rainfall} + \beta_6 \text{ critical rains} + \beta_7 \text{ shade trees} + \beta_8 \text{ pruning} + e \text{ ------------------------------ } (4.2)$$

where Y is the productivity of Robusta
α_0 is a constant or intercept
e is the unexplained error term

As seen in the above model, the square term was added to the age variable for two reasons. First, to understand more accurately the influence of age on the productivity of Robusta. Second, to know at what point the relationship between productivity and the independent variables tends to change.

From Table 4.1, it can be observed that the coefficient associated with the age of Robusta trees turned out to be positive, whereas its square term is negative. The coefficients associated with critical period rains and pruning of coffee trees turned out to be positive. On the other hand, the coefficient associated with elevation, area under coffee, average annual rainfall and shade trees per ha turned out to be negative. R^2 was found to be 0.79. Overall, the model was significant at the 1 per cent level of significance.

As seen from Table 4.1, the coefficient associated with age was found to be positive (33.90) and significant at the 1 per cent level of significance. The age square was found to be negative (–0.55) and significant. This meant that the productivity of Robusta initially responded positively with age, but later on, with increase in the age of coffee trees, it gradually declined. The maximum productivity was observed approximately between 20 and 30 years

Table 4.1 Factors affecting productivity of Robusta: an overall regression analysis (*Y* = kg/ha)

Variable	*Coefficient*	*'t' value*	*Level of significance*
Constant	772	4.97***	7.7E−06
Age of coffee trees (years)	33.90	6.06***	1.6E−07
Age square of coffee trees	−0.55	−6.57***	2.5E−08
Elevation (m)	−0.16	−1.91**	0.05
Area under coffee (ha)	−1.42	−1.75	0.18
Average annual rainfall (mm)	−0.01	−1.25	0.21
Critical period rainfall (mm)	2.49	5.58***	8.9E−07
Shade trees per ha	−0.005	−0.02	0.97
Pruning (₹/ha)	4.16	5.52***	1.1E−06
R^2	0.79		
F-value	25.15***		
No. of observations	60		

Source: Field survey.
***, ** and * significant at 0.01, 0.05 and 0.10 level, respectively.

in Robusta; however, the productivity gradually declined later, reflecting an inverse U-shaped relationship between age and productivity of Robusta (Figure 4.3).

Further, the inflection points[10] were estimated for the age variable in both Arabica and Robusta, as both the coefficients and their square terms turned out positive, wherein inflection points in Arabica and Robusta were 24 and 30 years, respectively, after which the productivity shows the sign of gradual decline (Table 4.17a).

It is seen from the results provided in Tables 4.16a and 4.1 that the productivity of coffee initially increases with the age of coffee trees, reaches a maximum, stabilizes, and with further increase in the age of coffee trees, the productivity gradually declines, which reflects an inverse 'U'-shaped relation between age and productivity of coffee trees, as shown in Figures 4.2 and 4.3. Therefore, the hypothesis that coffee productivity makes an inverse 'U' shape with respect to the age of coffee plantations in both Robusta and Arabica varieties can be accepted.

While studies have shown that as the coffee tree ages, the available nutrients are not efficiently utilized because the old trees have primaries and small heads, which adversely affect nutrient uptake (Onzima, 1996), while the field observations reveal that only a few fruits are produced on limited branches in the case of old trees. Besides, aged coffee trees etiolate and lack vigour, while the primaries with short internodes have limited space for the development of cherries. Under such circumstances, the application of fertilizers does not substantially increase the yield.

Besides, Hanisch et al., 2011 note the decline in potassium content with ageing coffee plantations. The old coffee trees do not respond well to nitrogen, phosphorus and potassium uptake, in contrast to the younger ones. On the other hand, younger coffee plants are characterized by finer roots which are restricted to the superficial layer of soil, while the older trees produce heavy roots which penetrate deep into the soil, adversely impacting nutrient uptake from the soil, thus impacting productivity.

Currently, about 10 per cent of surveyed coffee estates have crossed the economic bearing age, while the share of old-aged trees will significantly increase over time, adversely impacting the productivity of coffee in years to come, if replanting and rejuvenation are not carried out in the right time. The government has taken measures for replanting and rejuvenation in the 12th Five Year Plan to boost the productivity of Indian coffee. However, the region is dominated by small growers who are reluctant to remove the old plants and wait for the younger plants to attain maturity, which eventually requires a substantial amount of initial investment. The growers will have to forego their initial income during the pre-bearing period and fear the risks involved during that period wherein the growers' incomes might fail to meet the production costs.

In the case of Arabica, the coefficient associated with elevation was found to be positive. In contrast, the coefficient associated with Robusta was found to be negative (–0.16). This indicates that the productivity of Arabica increases with elevation and vice versa for Robusta.

One probable reason for lower productivity of Robusta at higher elevations could be that the higher elevation zones in Kodagu were earlier cardamom-growing belts. However, after the outbreak of mosaic virus disease (katte disease) in cardamom, resulting in huge losses, growers converted a large percentage of cardamom land into Robusta coffee, and they are still struggling with coffee cultivation, as this converted land was found to be less suitable for coffee cultivation (Deepika et al., 2013).

Generally, Robusta coffee requires warmer conditions for growth, unlike Arabica. The lower altitude (about 800 m) provides a warmer and harsher climate, wherein the Robusta can easily flourish, especially in the plain or flatter plantations. Likewise, Ramalho et al. (2014) note that Robusta coffee can thrive well at a lower altitude, characterized by higher temperatures.

Besides, the Robusta trees are more resistant to pests and diseases like White Stem Borer (*Xylotrechus quadripes*) and Coffee Leaf Rust (*Hemileia vastatrix*), partly due to a higher caffeine content (2.2 per cent caffeine in Robusta, while Arabica has 1.2 per cent). Further, Robusta can also withstand direct sunlight, with lower requirements for shade and scanty rains, making Robusta cultivation easier at lower altitudes. In contrast to this, the Robusta estates located at higher altitudes are characterized by heavier rainfall and frost conditions, especially the Robusta estates located in the Madikeri and Napoklu regions, resulting in fruit drop and wilt disease, which has an adverse influence on the productivity of Robusta. However, the productivity of Arabica was found to be higher at higher elevations, as it was associated with lower temperatures and lower incidence of coffee White Stem Borer (Murphy et al., 2008). Thus, from our analysis and the above discussions, the hypothesis that Arabica productivity has a direct relationship with the elevation of estates whereas in the case of Robusta the relationship is inverse can be accepted.

With the above analysis and given the regression results, it can be concluded that several factors influenced the productivity of coffee, while the age and elevation also influenced the productivity of both varieties of coffee. The average annual rainfall had a positive influence on the productivity of Arabica and a negative influence on Robusta. However, the coefficient associated with average annual rainfall was found to be insignificant in both the cases. Likewise, the critical period rains had a positive influence on the productivity of Robusta alone and were significant at the 1 per cent level. The Robusta crop is highly susceptible to drought and requires certain amounts of rain during the critical period. The insufficient critical rains in Robusta (blossom and backing showers) adversely affect the productivity of Robusta. The Robusta, when affected by drought, could limit the vegetative growth of the trees, indicating floral abnormalities followed by poor fruit setting. In contrast to this, the genetics of Arabica are more tolerant to drought and that the critical rains were insignificant in the case of Arabica. As seen in Table 4.16a, the coefficient associated with the shade trees was found to be positive and significant in the case of Arabica. Though the earlier literature had indicated a negative relation between shade and Arabica productivity, Kedlaya (2016) notes that the Arabica plantations yield better when grown under shade between 40 and 65 per cent, while Muschler (2001) observed the production of coffee trees grown under a homogenous shade of 30–50 per cent is equivalent to that of sun-grown coffee. In other words, the shade trees are beneficial to coffee production when grown at suboptimal biophysical conditions.

Besides, when coffee estates are provided with 'optimum shade', they in turn reduce the stress of coffee trees and buffer the high and low temperatures by 5°C. For instance, Barradas et al. (1986) document that in Mexican coffee plantations, the optimum shade can reduce the maximum temperature by 5.4°C, while Kirkpatrick (1935) note that in East African plantations, the shade could reduce the temperature by 5–6°C. However, the temperature in Kodagu varies between 16° and 39°C, while the optimum temperature requirement for Arabica is about 15–25°C. Thus, the optimum shade could reduce and provide ideal temperature requirements for Arabica to grow into healthy and vigorous plants. The optimum shade also favours uniform flowering of coffee, more abundant maturation of floral buds and flower synchronization, allowing uniform fruit maturation, besides enhancing photosynthesis, transpiration and stomata conductance. Undoubtedly, the reduced temperature due to optimum shade trees would bring advantages such as uniform flowering, higher recoveries of ripe berries, reduction in the dropping of ripe beans to a certain extent, which in turn increases the productivity of coffee. More importantly, optimum shade reduces White Stem Borer in Arabica.

However, in the case of Robusta, coefficients associated with shade trees were found to be negative but turned out to be insignificant. It was noticed during the field visits that in a few Robusta estates, the coffee wilt disease (caused by *Gibberella xylarioides*) was higher in the shaded coffee plantations, particularly in the Napoklu and Madikeri Robusta growing belts. Overall, the Robusta growers opined that lower shade enhances the productivity of Robusta, besides limiting the spread of Shot Hole Borer (*Xylosandrus compactus*), which develops under higher shade. Therefore, the growers either remove excess shade trees or maintain a thin shade. Though shot hole borer was not a major problem, it could reduce the productivity by less than 5 per cent (in estates comprising higher shade). Nevertheless, the Robusta estates in Kodagu, which require a relatively sparse shading, are now opening up the shade and replacing it with the silver oak in order to attain higher productivity and prefer to cultivate almost sun-grown coffee. Overall, the influence of shade trees on the productivity of Robusta was negative, but it turned out to be insignificant.

From the above discussions, it can be concluded that natural factors like rainfall and shade canopy do not always affect the productivity of coffee positively (shade trees positively affect the productivity of Arabica alone, while rainfall is found to be insignificant in both cases and critical rains are found positive only in Robusta). Therefore, the hypothesis that the productivity of both Arabica and Robusta has a positive relation with natural factors like rainfall (including critical rainfall) and shade canopy is partially refuted.

From Table 4.16a, it is evident that the White Stem Borer had a negative influence on the productivity of Arabica; viz. every 1 per cent increase in the uprooting of Arabica trees due to White Stem Borer resulted in a 7.66 per cent reduction in Arabica productivity. However, our field survey reveals that White Stem Borer (*Xylotrechus quadripes*) generally attacks Arabica (due to lower caffeine content with milder and juicier content in Arabica) and not Robusta. From the above regression analysis, the hypothesis that the White Stem Borer affects productivity negatively only in the case of Arabica while the Robusta variety is free from WSB incidence can be accepted.

4.10 Summary

In the current study, seven major factors (age of the coffee tree, elevation of estates, average annual rainfall, rainfall during the critical period, shade trees per ha, area under coffee and incidence of White Stem Borer) were taken into account to understand their influence on the productivity of coffee. The tabular and regression analysis revealed that among several factors, age had a significant influence on the productivity of both Arabica and Robusta varieties of coffee. The productivity of both Arabica and Robusta was found to be higher among the younger trees, remained stable for some time and further declined with the further age of coffee trees in both varieties, indicating an inverse 'U'-shaped relation between age and productivity of coffee trees. In both cases, the productivity was found to be low in estates with coffee trees above 40 years of age. However, about 3.33 per cent of Arabica and 11.66 per cent of the Robusta estates comprised trees that are over 40 years old. Therefore, rejuvenation of coffee trees older than 40 years is the need of the hour in order to attain the long-term economic viability of coffee estates.

The Arabica variety responded positively to elevation, whereas in the case of Robusta, productivity declined with elevation. Regarding rainfall, the Arabica variety responded positively to the average annual rainfall, whereas it was found to be negative in the case of Robusta. However, the coefficients associated with average annual rainfall were found insignificant in both the cases. Analyzing the effects of critical rains on the productivity of Robusta, it was found that the productivity of Robusta responded positively to critical rains. Therefore, during times of failure of critical rains, the Robusta plantations needed to be provided with mandatory sprinkler irrigation. Hence, the government should facilitate the growers to first install the sprinkler machinery and then apply for the subsidy, as it takes a very long time for clearance of formalities that could adversely impact productivity. In terms of shade trees per ha, the productivity of Arabica responded positively to the number of shade trees per ha. Although the productivity of coffee increased with an increase in area, the productivity of Robusta was found to decline above 31 ha; however, the coefficient was found to be insignificant. The productivity of coffee was higher in large-sized Arabica growers (26 to 50 ha), whereas minimum productivity was observed among the small growers (<2 ha). Overall, it can be concluded that among several factors influencing productivity, age had a major and significant influence on the productivity of coffee followed by other natural factors considered in the study.

Notes

1 Among the ten agro-ecological zones of Karnataka, Kodagu district comes under the hilly zone.
2 The Coffee Board has an annual re-plantation target of 5000 hectares as was set for the 12th five-year plan.
3 Quantity of fertilizer application, pruning, shade regulation, irrigation and plant protection chemical (PPC) application depend on the age of coffee trees.
4 The 2015 rain cycles have changed markedly, and heavy rains were received during the last three weeks of May.

5 The term 'critical period rains' has been used in the study for blossom and backing showers.
6 In case of failure of rainfall during the critical period, it should be supplemented by sprinkler irrigation.
7 The scientists across the globe could not precisely pinpoint the specific hormone responsible for floral induction but have narrowed it down to a light-mediated compound known as 'phytochrome', which, along with other growth regulators (auxin, gibberellic acid, acetic acid), induces flowering.
8 Botanical term used for a horizontal branch, upon which the leaves are arranged in one plant.
9 The coefficient associated with the square term of independent variable age was included in the model and was found to be significant. The square terms of other independent variables were not found to be significant and were therefore not included in the model.
10 The inflection point is the point at which the direction of the curve changes (either positive or negative).

References

Adugna, D.B. and Paul, C.S., 2011, Effects of shade on growth, production and quality of coffee (Coffea arabica) in Ethiopia. *Journal of Horticulture and Forestry,* **3**(11): 336–341.

Alvarado, Y., Pinto, S.L., Garcia, B.L. and Gaytan, J.F., 2002, Coffee yields and soil nutrients under the shades of Inga sp. vs. multiple species in Chiapas, Mexico. *Agroforestry Systems,* **54**(3): 215–224.

Ambinakudige, S. and Sathish, B.N., 2009, Comparing tree diversity and composition in coffee farms and sacred forests in the Western Ghats of India. *Biodiversity and Conservation,* **18**: 987–1000.

Avelino, J., Barboza, B., Davrieux, F. and Guyol, B., 2007, *Shade effects on sensory and chemical characteristics of coffee from very high-altitude plantations in Costs Rica.* Proceedings of International symposium on Multi-Strata Agroforestry Systems with Perennial crops: Making ecosystem services count for farmers, consumers, and the environment, Turrialba, Costa Rica, 17–21, September 2007.

Balaji, F., 2010, Decline in Indian coffee productivity levels alarming: Upasi. *Headlines India,* September 14. Available at http://headlinesindia.mapsofindia.com/business-news/coffee/decline-in-indian-coffee-productivity-levels-alarming-upasi-63203.html

Bandeira, F.P., Martorell, C., Meave, J.A. and Caballero., 2005, The role of rustic coffee plantations in the conservation of wild tree diversity in the Chinantec region of Mexico. *Biodiversity and Conservation,* **14**: 1225–1240.

Barradas, V.L. and Fanjul, L., 1986, Microclimatic characterization of shaded and open-grown coffee (Coffea arabica L.) plantations in Mexico. *Agricultural and Forest Meteorology,* **38**: 101–112.

Benjamin, Dwayne and Deaton, Angus, 1993, Household welfare and the pricing of cocoa and coffee in Cote d' Ivoire: lessons from the living standards surveys. *The World Bank economic review,* 7(3): 293–318.

Chengappa, P.G., Rich, K.M., Muniyappa, A., Yadava, C.G., Ganashruthi, M.K., Pradeepa Babu, B.N., Shubha, Y.C. and Rich, M., 2014a. *Sustainability coffee certification in India*. Perceptions and Practices. NUPI Working Paper 830, Norwegian Institute of International Affair.

Coste, R., 1992, *Coffee – The plant and the product*. MacMillan Press, London.

Covre, A.M., Fabio, L.P., Robson, B., Braun, H. and Ronchi, C.P., 2016, Vegetative growth of Conilon coffee plants under two water conditions in the Atlantic regions of Bahia state, Brazil. *Acta Scientiarum: Agronomy,* **38**: 535–545.

Craparo, A.C.W., 2012, Influence of climate change on coffee yield and production in East Africa: 1961–2009. In *24th International Conference on Coffee Science*, 11–16 November 2012 (pp. 358), Programs and Abstract, San Jose, Costa Rica: ASIC.

DaMatta, F.M. and Cochico Ramalho, J.D., 2006, Impacts of drought and temperature stress on coffee physiology and production: A review. *Brazilian Journal of Plant Physiology,* **18**(1): 55–81.

Deepika, M.G. and Jyotishi, A., 2013, *Assessing risk and risk mitigation strategies of small coffee growers: A study of Kodagu district in Karnataka*. NRPPD Discussion Paper No. 21, Centre For Development Studies, Thiruvanathapuram.

Duke, J.A., 1983, *Handbook of energy crops*. Published online. www.hort.purdue .edu/newcrop/duke_energy/ Coffea_arabica.html.7p

Evizal, R., Sugiatno, S., Prasmatiwi, F.E. and Nurmayasari, I., 2016, Shade tree species diversity and coffee productivity in Sumberjaya, West Lampung, Indonesia. *Biodiversitas,* **17**: 234–240.

Garcia, C.A., Bhagwat, S.A., Ghazoul, J., Nath, C.D., Konerira, M.N., Cheppudira, K.G., Raghuramulu, Y., Nasi, R. and Vaast, P., 2010, Biodiversity conservation in agricultural landscapes: Challenges and opportunities of coffee agroforests in the Western Ghats, India. *Conservation Biology,* **24**(2): 479–488.

Gopakumar, C.S., 2011, *Impacts of climate variability on Agriculture in Kerala*. Doctoral dissertation. Cochin University of Science and Technology, Cochin.

Guyot, B., Manez, J.C., Perriot, J.J., Giron, J. and Villain, L.,1996, Influence de l'altitude et de l'ombrage sur la qualité des cafés arabica. *Plant Research and Development Journal*, **3**: 272–280.

Hajian, F.Z., Gonthier, D.J., Marin, L., Iverson, A.L. and Perfecto, I., 2014, Changes in species diversity of arboreal spiders in Mexican coffee agroecosystems: Untangling the web of local and landscape influences driving diversity. *PeerJ,* **2**: e623.

Hanisch, S., Dara, Z., Brinkmann, K. and Buerkert, A., 2011, Soil fertility and nutrient status of traditional Gayo coffee agroforestry systems in the Takengon region, Aceh Province, Indonesia. *Journal of Agriculture and Rural Development in the Tropics and Subtropics*, **112**(2): 87–100.

Jaramillo, J., Mamoudou, S., Eric, M., Olaye, A.C., Jaramillo, A., Mukabana, J., Maina, J., Gathara, S. and Borgemeister, C., 2013, Climate change or urbanization? Impacts on a traditional coffee production system in East Africa over the last 80 years. *PLoS One*, **8**(1): 1–10.

Jassogne, L., Laderach, P. and Van Asten, P., 2013, The impact of climate change on coffee in Uganda. Lessons from a case study in the Rwenzori Mountains. *Oxfam Policy and Practice: Climate Change and Resilience*, **9**(1): 51–66.

Jezeer, R., Verweij, P., Santos, M. and René, G.A., 2017, Shaded coffee and cocoa – Double dividend for biodiversity and small-scale farmers. *Ecological Economics*, **140**: 136–145.

Joseph, K.J. and George, P.S., 2010, *Structural infirmities in Indian Plantation Sector: Natural rubber and spices*. Report submitted to the Ministry of Commerce, NRPPD, and Centre for Development Studies, Trivandrum.

Indian Coffee Board. 2017, Database on Coffee, March 2017. Available at www.indiacoffee.org

International Coffee Organization (ICO), 2014, *World Coffee Trade (1963-2013): A review of the Markets, Challenges and Opportunities Facing the sector*. 112th Session, ICO, London.

Kedlaya, Gana, 2016, Bitter coffee. Available at http://www.deccanherald.com/content/570011/bitter-coffee-this.html

Kirkpatrick, T.W., 1935, *The climate and Eco-climates of coffee plantations*. Amani Memoirs, No. 28, Crown Agents, London, 6pp.

Kodagu District Census Handbook,2011, Government of Karnataka. Available at https://cdn.s3waas.gov.in/s3c8ed21db4f678f3b13b9d5ee16489088/uploads/2019/06/2019061365.pdf

Kumar, Parmod and Sharma, Anil, 2006, Perennial crop supply response functions: The case of Indian rubber, tea and coffee. *Indian Journal of Agricultural Economics*, **61**(4): 613–646.

Mitchell, H.W. (1985) Cultivation and harvesting of the Arabica coffee TREE. In: Clarke, R.J. and Macrae, R., (Eds.,) *Coffee agronomy*, Vol. 4, Booker Agriculture International Limited, Westport Publishing Company, London, 163-178 pp.

Moguel, P. and Toledo, V.M., 1999, Biodiversity conservation in traditional coffee systems of Mexico. *Conservation Biology,* **13**: 11–2l.

Morais, H., Caramori, P.H., Ribeiro, A.M. de Arruda., Gomes, J.C. and Koguishi, M.S., 2006, Microclimatic characterization and productivity of coffee plants grown under shade of pigeon pea in Southern Brazil. *Pesquisa Agropecuaria Brasileira,* **41**(5): 763–770.

Msuya, A.M., 2013, *Impact of climate variability on coffee production and farmers coping and adaptation strategies in highlands of Kigoma district*, Tanzania. M.A (Rural Development) thesis, Sokoine University of Agriculture, Morogoro, Tanzania.

Murphy, T.S., Phiri, N.A., Sreedharan, K., Kutywayo, D. and Chanika, C., 2008, *Integrated stem borer management in smallholder coffee farms in India, Malawi and Zimbabwe*. Final Technical Report. CFC/CABI/ICO/CRS/CB.

Muschler, R.G., 2001, Shade improves coffee quality in a sub-optimal coffee-zone of Costa Rica. *Agroforestry Systems*, **51**(2): 131–139.

Ngabitsinze, J.C., Mukashema, A., Ikirezi, M. and Niyitanga, F., 2011, *Planning and costing adaptation of perennial crop systems to climate change: Coffee and banana in Rwanda*. International Institute for Environment and Development (IIED), London, UK.

Onzima, R.J., Lukwago, G., Wetala, M.P.E. and Wejuli, M., 1996, Effects and economics of nitrogen fertilizers on the changing cycle of old robusta coffee in Lake Victoria cresent: In improving coffee management systems. *African Crop Science*, **2**: 88–92.

Potvin, C., Owen, C.T., Melzi, S. and Beaucage, P., 2005, Biodiversity and modernization in four coffee-producing villages of Mexico. *Ecology and Society,* **10**(1): 18.

Ramalho, J.C., DaMatta, F.M., Rodrigues, A.P., Scotti Campos, P., Pais, I., Batista-Santos, P. and Leitao, A.E., 2014. Cold impact and acclimation response of Coffea spp. plants. *Theoretical and Experimental Plant Physiology*, **26**(1): 5–18.

Ricci, M.S.F., Rouws, J.R.C. and Oliveira, N.G., 2011, Vegetative and productive aspects of organically grown coffee cultivars under shaded and unshaded systems. *Scientific Agriculture*, **68**(4): 424–430.

Santos, K.M., Moura, M.F.V., Azevedo, F.G., Lima, K.M.G., Raimundo, I.M. and Pasquini, C., 2012, Classification of Brazilian coffee using near-infrared spectroscopy and multivariate calibration. *Analytical Letters*, **45**: 774–781.

Silva, E.A., Mazzafera, P., Brunini, O., Sakai, E., Arruda, F.B., Mattoso, L.H.C., Carvalho, C.R.L. and Pires, R.C.M., 2005, The influence of water management and environmental conditions of the chemical composition and beverage quality of coffee beans. *Brazilian Journal of Plant Physiology*, **17**(2): 229–238.

Somporn, C., Kamtuo, A., Theerakulpisut, P. and Siriamornpun, S., 2012, Effect of shading on yield, sugar content, phenolic acids and antioxidant property of coffee beans (Coffeaarabica L. cv Catimor) harvested from north-eastern Thailand. *Journal of the Science of Food and Agriculture*, **92**: 1956–1963.

Sridevi, V. and Giridhar, P., 2013, Influence of altitude variation on trigonelline content during ontogeny of coffea canephora fruit. *Journal of Food Studies*, **2**(1): 62–74.

Shrinidhi, A.S. and Lele, S., 2001, *Forest tenure regimes in the Western Ghats: A compendium*. Institute for Social and Economic Change, Bangalore, **90**: 56.

Thorpe, J. and Fennell, S., 2012, Climate change risks and supply chain responsibility. How should companies respond when extreme weather affects small-scale producers in their supply chain? *Oxfam Policy and Practice*, **9**(1): 39–62.

Titus, A. and Pereira, G.N., 2011, The fine art of irrigation in Robusta coffee Plantations. Available at https://ecofriendlycoffee.org/the-fine-art-of-irrigation-in-robusta-coffee-plantations

Tolessa, K., Rademaker, M., De Baets, B. and Boeckx, P., 2016, Prediction of specialty coffee cup quality based on near infrared spectra of green coffee beans. *Talanta*, **150**: 367–374.

Vaast, P., Bertrand, B., Perriot, J.J., Guyot, B. and Genard, M., 2006, Fruit thinning and shade improve bean characteristics and beverage quality of coffee (Coffea arabica L.) under optimal conditions. *Journal of the Science of Food and Agriculture*, **86**(2): 197–204.

Chapter 5

The coffee White Stem Borer (*Xylotrechus quadripes*) and its impact on Arabica variety of coffee

In the recent past, there has been a radical shift from Arabica to Robusta coffee cultivation in the country. The dimension of the shift is to such an extent that the much-demanded Arabica variety is under 'high risk of extinction' (Davis et al., 2012). Nevertheless, Arabica coffee is considered to have a higher cup quality than Robusta. Factually, the lesser caffeine content (0.9–1.7 per cent) of Arabica, high bean quality, milder taste and fine cup qualities reflect the superior quality of Arabica over Robusta (Belachew, 2003) thus making Arabica variety exclusive for nearly all 'speciality coffee' (Davis et al., 2012). Given the above facts and the premium for better quality, the price of Arabica remains much higher as compared to Robusta (Figure 5.1).

The productivity of Indian Arabica is much lower than that of Robusta (Figure 5.2). The earlier literature states the reasons for the decline in Arabica owing to pests and diseases, changing weather and climatic conditions and higher costs of management. Among the various pest attacks, the major one is the coffee White Stem Borer (*Xylotrechus quadripes*), considered to be the major reason for the decline in Arabica productivity in the recent period and ranked to be the number one major production constraint in *Coffea arabica* (Murphy et al., 2008).

5.1 White Stem Borer and its economic impacts on Arabica

The widespread attack of this pest over the years has led to the subsequent uprooting of plants, thereby reducing plant population and productivity, which resulted in cumulative revenue loss to the growers (Venkatesh et al., 2012). The reduction in the yield of Arabica due to Stem Borer, coupled with high risks and increasing costs of Arabica cultivation, was the major cause for the shift towards Robusta, which has relatively better yield and lower cost of cultivation compared to Arabica (Deepika et al., 2013).

In India, the average estimated loss of productivity in individual estates caused by Stem Borer was above 25 per cent, which aggregated to a substantial annual economic loss of ₹450 million to the nation (Murphy et al., 2008).

DOI: 10.4324/9781003541721-5

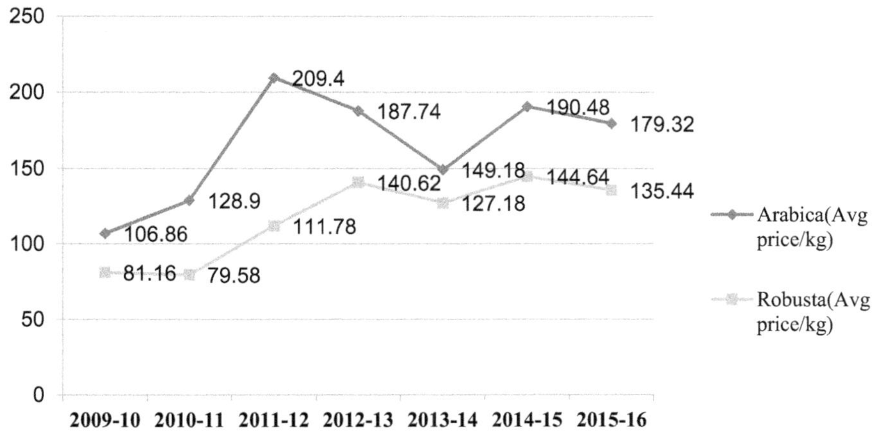

Figure 5.1 Average price of Arabica and Robusta over the years (₹/kg).
Source: Indian Coffee Board (2017).

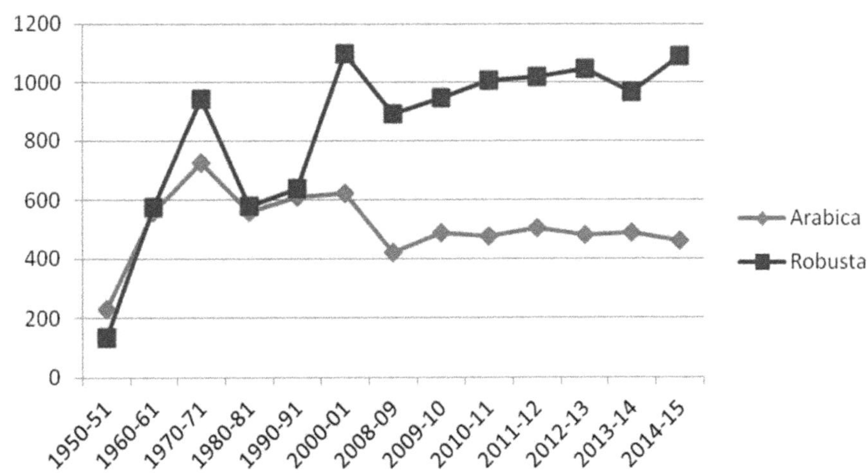

Figure 5.2 Productivity of Arabica and Robusta over years (at all India level).
Source: Indian Coffee Board, 2017.

As per Subramaniam (1934), the estimated average productivity loss in heavily infested plantations could vary between 17.80 per cent and 20 per cent, but if integrated management practices were followed, losses could vary between 2.3 per cent and 12.50 per cent. Radhakrishnan (1987) estimated that the removal of stem borer-infested plants in India, even at the rate of one plant per hectare, could cost a total loss to the nation of about US $642,585 per annum, while Venkatesha et al. (2012) estimated the annual economic loss to be as high as US$ 17.5–26 million. As per the report of Hall (2006), *Xylotrechus quadripes* could cause an annual loss of US$ 40 million for uprooting, replacement and production loss, thereby destroying over nine million trees annually. Likewise, a recent report estimated that the White Stem Borer (WSB) could cause a loss of about half a million Arabica plants per year, for which the estimated loss was 1500 tonnes per annum for the country (Businessline, 2015).

In Karnataka during 2014, nearly 3200 hectares of coffee plantations were affected by White Stem Borer. In other words, about 50 per cent of Arabica estates in Karnataka were infested by WSB, while in Kodagu district alone, significant yield loss of 35 per cent occurred each year due to stem borer. Unable to bear the economic burden of White Stem Borer, several estates were left neglected, without the application of fertilizers, insecticides or any plant protection chemicals. Given the perennial nature of coffee, crop losses could even reach the extent of 93.60 per cent in plantations under poor management (Basavaraj et al., 2005), while the adverse impact of this pest could last up to 5 years (Murphy et al., 2008). This has, in turn, majorly impacted the marginal and small growers whose livelihoods depend on the production and income from coffee farming. The Coffee Board has also opined that White Stem Borer remains a major threat in the formulation of policies for the promotion of Arabica.

5.2 Opinion of growers about shifting and willingness to shift towards Robusta

Table 5.1a summarizes selected households' willingness to shift towards Robusta or not willing to shift in accordance with the size of land holdings. About 68.33 per cent of the interviewed growers were replanting (partially) Robusta plants in place of Arabica. Among them, about 38.33 per cent belonged to the category of small growers, while about 30 per cent were large holders. About 21.67 per cent of growers were willing to shift, but unable to shift towards Robusta, wherein 18.33 per cent were small growers, while only 3.34 per cent were large growers. On the other hand, about 10 per cent of the interviewed growers continued with Arabica cultivation despite several constraints and were not willing to shift towards Robusta.

An important part of this survey was to understand the various factors responsible for this shift. A five-point Likert scale (strongly agree received a ranking of 5, while strongly disagree received a ranking of 1) was used to understand the factors responsible for the shift, non-shift and willingness to, but unable to, shift among the Arabica growers.

5.2.1 Opinion of Arabica growers to shift towards Robusta

The growers' opinion is summarized in Figure 5.3. It was evident from the results that about 68.33 per cent of growers were in the process of shifting

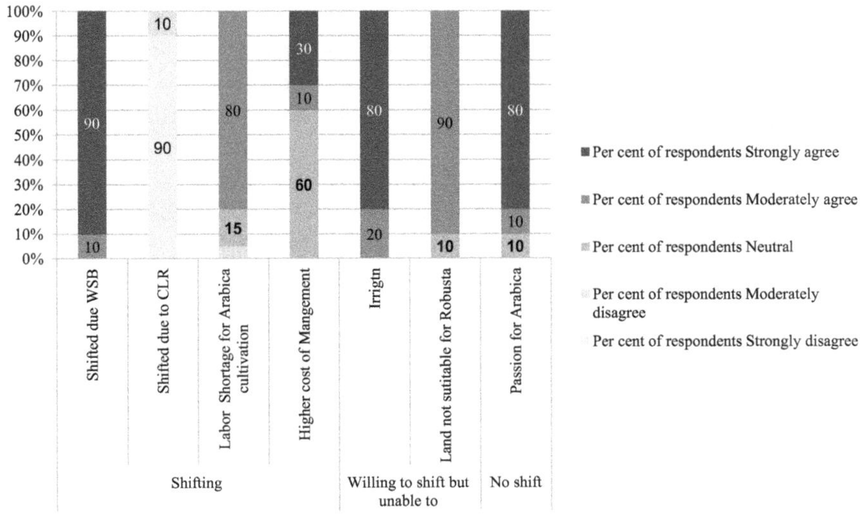

Figure 5.3 Opinion of Arabica growers on shifting towards Robusta, willingness to shift, but unable to, shift and non-shift towards Robusta.

Source: Field survey.

towards Robusta. As expected, White Stem Borer (*Xylotrechus quadripes*) was rated high by the growers as the number one and major cause for the shift, ranked as the major production constraint in *Coffea arabica*. Among the growers shifting towards Robusta, about 90 per cent strongly agreed that the heavy infestation of White Stem Borer in their estates with resultant heavy crop loss prompted them to shift towards Robusta (Figure 5.3). In our survey, the Arabica growers responded that the average annual yield loss due to WSB in individual estates was about 6.45 per cent, while the average uprooting was 150 plants per ha. The field visits in Kodagu revealed that higher incidences were observed in S.795 and Sln.6 varieties of Arabica. Nevertheless, the lower incidence of borer attack in the Sln.5A variety of Arabica was due to its inbuilt genetic tolerance for stem borer. However, none of the respondents in our survey reported the Sln.5A variety of Arabica cultivated in their estates. Therefore, growers in Kodagu should be encouraged to cultivate the Sln.5A variety of Arabica.

Though the Arabica variety was also more susceptible to a fungal disease, Coffee Leaf Rust (CLR), caused by *Hemileia vastatrix*, the surveyed growers' opinion was that currently CLR was under control due to the regular application of Bordeaux mixture. Likewise, Narayana (2013) found that the occurrence of CLR was decreasing in Kodagu. Overall, most of the growers (about 90 per cent), who were shifting towards Robusta, strongly observed that CLR was not the reason for the shift. Further, about 80 per cent of interviewed growers moderately asserted that the shortage of a 'skilled' labor force had also paralyzed the Arabica plantation in Kodagu, while about 15 per cent of growers took a neutral stand in this regard. The severe shortage of skilled labour force, despite the hike in wage rates, had also prompted the Arabica growers to search for alternative solutions by substituting Robusta in place of Arabica as the Robusta cultivation required less labour force (Figure 5.3). About 30 per cent of growers strongly agreed that the higher cost incurred in Arabica cultivation in comparison to Robusta induced them to shift towards Robusta, which had fairly lower production costs. About 60 per cent of interviewed growers took a neutral stand in this regard (Figure 5.3). Nonetheless, whatever may be the factors responsible for the shift, it is certainly a concern for the entire coffee industry, as they fear its extinction in near future. Not only in Kodagu, even at the inter-continental level, scientists estimate global extinction of Arabica within the next 70 years (Plat, 2012).

5.2.2 *Opinions of growers regarding willing to shift, but unable to shift, towards robusta*

As discussed earlier, about 21.67 per cent of the interviewed growers were willing to shift to Robusta but unable to do so. Among them about 80 per cent of the growers strongly agreed that the lack of irrigation facilities in their estates constrained them from shifting to Robusta, as water requirements for Robusta were very high. In contrast, Arabica, being drought-tolerant crop, required less irrigation than Robusta (Titus, 2011), as our household survey indicated that bore wells and tanks were the major sources of irrigation in the study region. However, about 18 per cent of growers largely relied on rainfall (Figure 5.4) having the lack of alternate sources of irrigation. The lack of irrigation was specifically predominant in the northeastern belt of Somwarpet, where the mean annual rainfall was found to be scanty, around 65 inches per

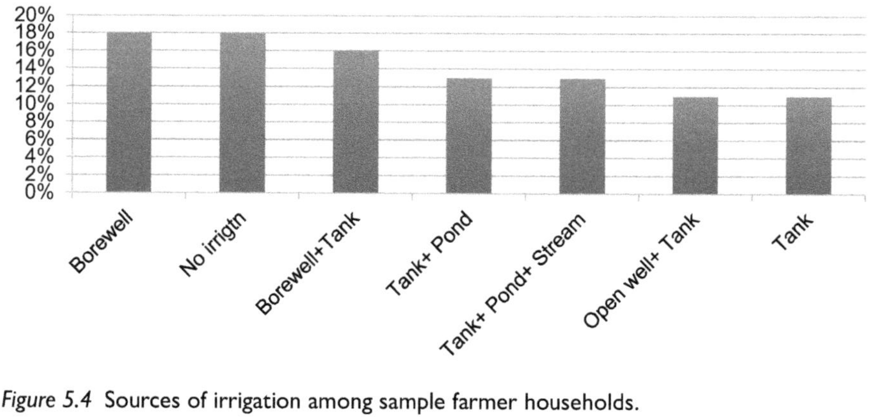

Figure 5.4 Sources of irrigation among sample farmer households.

Source: Field survey.

year. Though the Arabica growers in this region had constructed tanks for the purpose of irrigating coffee, but because of scanty rainfall, water storage remained below the desired level, disabling them from having sufficient irrigation facilities through tanks. Furthermore, the Arabica growers who had no tanks sunk bore wells, thereby exploiting the groundwater table. Apart from this, about 90 per cent of them moderately agreed that their land was unsuitable for Robusta cultivation.

5.2.3 Opinions of growers not willing to shift towards robusta

About 10 per cent of the interviewed growers were not willing to shift towards Robusta, and they continued cultivating Arabica despite several constraints. Among them, around 80 per cent strongly believed that Arabica coffee had better future prospects, thereby they wanted to save Arabica for future generations and it prompted them to protect Arabica from the risk of becoming extinct. One large grower indicated that he directly exported his Arabica coffee to the United Kingdom for which he received an extra premium; therefore, he was more assertive for Arabica cultivation.

Furthermore, the growers' opinion that the cultivation of Arabica helped them to preserve the natural shade trees and biodiversity as Arabica plants required higher shade requisites and were cultivated under the dense cover of indigenous shade trees (Garcia et al., 2010), while Arabica estates maintained a higher number of shade trees than Robusta estates (Chengappa et al., 2014a). In contrast, the gradual shifts from Arabica to Robusta have also reduced the shade density in coffee estates of the Western Ghats (Leroy, 2011). This reduction in shade trees might also adversely impact the rainfall patterns in Kodagu in the future years.

The growers who were not willing to shift towards Robusta had replanted the Chandragiri variety of Arabica in place of the uprooted varieties. The respondents' opinion was that Chandragiri was resistant to WSB, and the whole stem, covered with thick branches of leaves, provided higher shade cover and thus did not provide 'optimum open space and sunlight' required by WSB, thereby acting as a physical barrier to borer attacks. Moreover, the Chandragiri variety was resistant to both Leaf Rust and White Stem Borer. Henceforth, they pointed out that Chandragiri was relatively a better variety in Kodagu. These growers were highly passionate about Arabica cultivation and showed no signs of shifting towards Robusta.

5.3 Factors responsible for White Stem Borer (Xylotrechus quadripes) attack in Arabica coffee

The surveyed growers noted that they observed WSB right from the inception of their estates and that incidence was found to be increasing over the years, while currently they faced severe incidence of borer attack. There were several factors responsible for the borer attack which could be classified based on the environment and management practices adopted in individual estates. Among the environmental factors, about 47 per cent of growers ranked the rise in temperature as the number one cause for borer attacks, 35 per cent of growers opined erratic rainfall and 18 per cent of growers ranked elevation as the major environmental factors responsible for the borer outbreak. The management practices included the flight of WSB from neglected neighbouring estates (30 per cent), improper shade management (24 per cent) and ineffective control measures (24 per cent), which were ranked top three by the growers. The other factors included lack of skilled labour for the detection of WSB (15 per cent), nearby paddy fields (3 per cent), lack of host plants (3 per cent) and lack of research (3 per cent) on WSB control.

Taking into account the field observations, incidences of borer were classified on the basis of elevation and rainfall. For instance, the surveyed estates <1000 metres were classified as low elevation (LE) and >1000 metres as high elevation (HE) estates. Accordingly, the mean annual rainfall of 1397–1524 mm, 1524–2032 mm and 2032–2413 mm were classified as low rainfall (LR), medium rainfall (MR), and high rainfall (HR) regions. In accordance, estates were further classified into six categories, viz. lower elevation and lower rainfall (LE + LR), lower elevation and medium rainfall (LE + MR), lower elevation and high rainfall (LE + HR), high elevation and low rainfall (HE + LR), high elevation and medium rainfall (HE + MR) and High elevation and high rainfall (HE + HR). The incidence of the White Stem Borer was classified according to the number of plants uprooted per hectare. If the uprooting of plants was less than 25 per ha, they were classified as low incidence; similarly, uprooting 25–62 plants/ha, 62–125 plants/ha, 125–250 plants/ha, 250–375 plants/ha and >375 plants/ha were classified as medium, high, very high incidence, critical and alarming rate, respectively (Table 5.2a).

The incidences of White Stem Borer were very high in estates located at low elevation and low rainfall (LE + LR) estates. Among the 60 estates surveyed, about 60 per cent fell under the LE + LR category, out of which 20 per cent suffered very high incidences of borer attack, wherein the uprooting ranged from 125 to 250 plants per ha in a year. Few estates experienced low incidences of WSB, in which fewer than 25 plants per ha were uprooted. Meanwhile, about 13 per cent of estates experienced medium incidences of pest attack, where the uprooting varied between 25 and 62 plants per ha. However, one estate visited showed incidences at an alarming rate wherein the uprooting was more than 375 plants per ha. About 22 per cent of surveyed estates fell under the lower elevation and medium rainfall category, and among these, two estates suffered very high incidences and one estate experienced a borer attack at an alarming rate. Under the third category of lower elevation and high rainfall region, about 3 per cent of estates had medium borer incidences, and 2 per cent had high incidences. About 10 per cent of estates fell under the category of high elevation (HE) and low rainfall (LR), whereby 3 per cent of estates experienced medium and very high incidences. Under the fifth category of

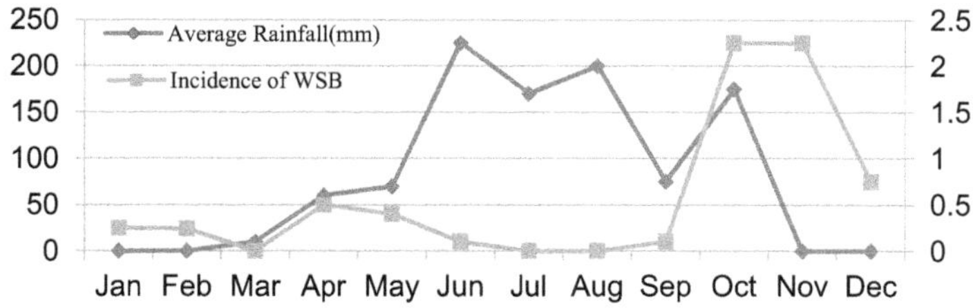

Figure 5.5 Rainfall (mm) and incidence of coffee White Stem Borer at CRS, Chetalli.

Source: Kumar P.K.V., 2009.

high elevation and medium rainfall, about 2 per cent estates experienced each of high and very high incidences. It was interesting to note that White Stem Borer incidence was absent in high elevation and high rainfall zones.

The variations in climatic conditions, viz. increased temperature, erratic rainfall and lower elevations, provided favourable conditions for the breeding of the White Stem Borer. In contrast, heavy rains, low temperatures and higher elevations control the spread of the borer. As per the field trials conducted by the Coffee Board inChetalli, Kodagu, on the flight patterns of the White Stem Borer in Chetalli (Kodagu) from 2002 to 2006 for a period of five years, an inverse relationship between rainfall and the incidence of the White Stem Borer was indicated (Figure 5.5). On the other hand, the dry spells during April, coupled with higher summer temperatures and delayed blossom showers, aggravate the breeding and spread of WSB. Likewise, our survey also indicated that WSB had contrasting effects with elevation and rainfall.

5.4 Regression analysis of factors influencing uprooting of Arabica per Ha

The uprooting of Arabica trees due to WSB was regressed on elevation (metres), age of Arabica trees (years), area under coffee (ha), rainfall (mm), shade trees/ha (number) and expenditure on pesticides (₹/ha). The functional form of the model is given below.

$\ln Y$ (number of trees uprooted) = $\alpha_0 + \beta_1 \ln$ (area under coffee) + $\beta_2 \ln$ (elevation) + $\beta_3 \ln$ (actual rainfall) + $\beta_4 \ln$ (shade trees per ha) + $\beta_5 \ln$ (expenditure on pesticides) + $\beta_6 d_1 + \beta_7 d_2 + e$ --(5.1)

where Y is the number of Arabica trees uprooted per ha

α is a constant or intercept
$d_1 = 1$, if age is 20–30 years; 0 otherwise
$d_2 = 1$ if age is more than 30 years, 0 otherwise
e is the unexplained error term

Firstly, the area under coffee (ha) was taken into account to understand if the incidence of WSB attack was more severe on small- or large-sized holdings. Secondly, the inclusion of elevation was intended to capture the impact of altitude on the incidence of WSB, viz. whether or not the probability of uprooting was higher in estates located at higher altitudes or the ones located at lower altitudes. Given the wide variation in rainfall distribution among the surveyed estates, the annual rainfall (mm) factor was included in the model to understand the influence of rainfall on the uprooting of Arabica due to WSB attack. The shade trees were highly recommended by the Coffee Board to control WSB. Now to what extent the shade trees control the White Stem Borer was the key question. In-order to answer this question, the number of shade trees per ha in the Arabica estates was taken into account. Besides, the spraying of pesticides (Chloropyrifos and Lindane) was a major production practice followed by the growers to control WSB. However, the growers expressed that the use of chemical pesticides did not help them control borer attacks. So, does the spraying of pesticides actually help the growers control the incidence of WSB? Therefore, the expenses incurred on pesticides were included in the model. Lastly, the study intends to understand the age group of Arabica trees that witness a higher incidence of borer attacks. Therefore, dummies on the age of trees between 20 and 30 years and above 30 years were included in the model.

From the regression results presented in Table 5.3a, it is seen that the value of R^2 was found to be 0.71. It can be observed from the table that every 1 per cent increase in elevation, the uprooting of Arabica decreased by 8.36 per cent. In other words, the incidence of WSB was lower in estates located at a higher elevation. Similarly, Murphy et al. (2008) have documented that the incidence of WSB decreased with an increase in the elevation and that the highest incidence was identified in estates located below 1036 m.a.s.l.

The incidence of WSB decreased with an increase in the number of shade trees per ha (Table 5.3a). For every 1 per cent increase in the number of shade trees, the uprooting of Arabica decreased by 0.41 per cent. The Arabica growers in Kodagu adopted a two-tier mixed shade pattern in their estates, with Dadap (*Erythrina lithosperma*) as a lower canopy of shade and the permanent trees like timber species (*Dalbergia latifolia*), Jack fruit (*Artocarpus integrifolia*) and Mahogany (*Cedrela toona*). These shade trees do not harbour the WSB and thereby the estates with a greater number of shade trees were less prone to borer attacks.

The expenditure on pesticides (₹/ha) was used as one of the factors in the model. The associated coefficient was negative and found to be significant at the 5 per cent level of significance. Thus, the expenses on pesticides (Chloropyrifos, Lindane, Contof) reduced the incidences of WSB in individual estates. These results were in contrast to the growers' belief that the use of chemical pesticides did not help them control the borer attack. It was found that every 1 per cent increase in the expenses on pesticides, the incidence of WSB reduced by 0.51 per cent.

In order to assess the impact of the incidence of the White Stem Borer on the physiological age of Arabica trees, we added three dummy variables as categorical predictors, viz. Dummy variable 1 = age of plants ≤ 20 years, Dummy variable 2 = age of plants between 20 and 30 years and Dummy variable 3 = age of plants >30 years. The major question was to determine whether there exists any relationship between different 'age groups' and the overall uprooting of the Arabica plants. The results revealed that the incidence of the White Stem Borer was higher in the plants with the age group between 20 and 30 years, while the incidence was lower in older trees above 30 years of age. Likewise, the younger Arabica plants of 4–5 years old, which have just entered the bearing period, are heavily attacked by the borer, while the older trees are inclined to live with borer attacks and manifest the signs of borer attack only under moisture or bearing stress. Generally, the old estates under good agronomic practices could easily sustain borer attack without any adverse impact on yield (Murphy et al., 2008). In the case of young plants, the larva bores into the stem, resulting in the death of young plants. Though the older Arabica plants survive the borer attack, there has been a profound reduction in the yield, and they were more susceptible to diseases. In contrast, studies have also indicated that borer incidence was high in older plants (8.3 per cent) and the incidence was only 0.7 per cent in very young plants (Shylesha and Veeresh, 1995). However, the coefficients associated with the area under coffee (ha) and rainfall (mm) were found to be insignificant in the model.

5.5 Costs incurred for WSB control

The most common control measure adopted by growers included chemical control. About 24 per cent of the surveyed growers pointed out that chemicals like Chloropyriphos and Lindane were ineffective in controlling borers, while Gammexane was highly effective in borer control and that WSB infestation increased after BHC was banned. The coffee-importing countries have also imposed stringent control over minimal residual Lindane application on coffee, thereby limiting growers on Lindane application.

The cultural methods included borer tracing, detection, handpicking of beetles, uprooting and burning the infested plants (Duffy, 1968). It was evident from Table 5.4a that the cultural costs increased with the incidence of attack, ranging between ₹670 per ha and ₹7500 per ha.

About 56 per cent of the growers practised Chloropyrifos spraying, for which cost measures were about ₹7425 per hectare, out of which 72.63 per cent was material cost, while 27.37 per cent was labor costs (Table 5.5a). About 23 per cent of growers applied Lindane, which cost about ₹8372 per hectare. Along with this, the combination of Lindane and Contof as well as Lindane and Neem had costs of about ₹9000 and ₹8750 per hectare, respectively. About 10.90 per cent of the growers applied a mixture of Chloropyrifos and Lime for which the estimated cost was about ₹8038 per hectare. On the other hand, scientists observe that the Chlorpyriphos and Lindane used for WSB control were highly toxic and that the blanket application of these insecticides by growers filtrate the soil and groundwater, including wells and rivers, making the water unfit for drinking purposes. However, about 5.10 per cent of the growers belonged to the non-adoption category, relying on cultural methods only for WSB control.

5.6 Institutional support for control of White Stem Borer (WSB)

The use of pheromone traps for WSB control was first demonstrated by Venkatesha in 1986. However, growers find it expensive since 1 hectare requires about 25 traps (Jayarama, 2009). To tackle this issue, the Coffee Board provided pheromone traps at subsidized rates of 50 per cent. From our survey, it was found that about 26 per cent of growers had installed pheromone traps at a height of 2.4 metres, set up at a distance of 20 metres apart. These were provided by the Coffee Board at a subsidized rate of ₹50 per trap. However, all the surveyed growers pointed out that the use of pheromone traps was not very effective in controlling the borer. They also opined that along with WSB, it also trapped other insects, in turn disturbing the food web. In general, the response of growers to pheromone trap technology was not encouraging, and growers were no longer interested in continuing this practice.

The Pest and Disease Act, 1968 implemented strict rules to control stem borer, wherein Kodagu, Chickmagalur and Hassan districts were declared as 'White Stem Borer affected areas' for a period of five years from the notification dated 13 July 2001, and Coffee White Stem Borer was declared a serious pest under section 3 of the Karnataka Pests and Disease Act. Under section 19 of this Act, the extension officers of the Coffee Board were given the responsibility of inspecting stem borer affected Arabica estates. Meanwhile, the Board had also directed the growers to uproot and burn the infested plants so as to prevent the further spread of the disease. However, the borers made entry through the neighbouring, neglected estates. In such cases, the Act provided powers to Coffee Board inspecting officers to issue notices against the growers of neglected estates. If the growers failed to take preventive measures, the officer had every right to take any remedial action to prevent borers in such neglected estates. The costs incurred by the officers were in turn recovered from the growers as arrears of land revenue. Besides, the Act had also restricted the movement of infested plants, soil or manure from one place to another (Murphy et al., 2008). Earlier, the Board had also encouraged the growers to collect the borers and paid them ₹1–1.5, while the uprooted plants were purchased by the tobacco growers of the neighbouring district at the rate of ₹4 per plant (Ambinakudige, 2006).

Furthermore, to improve the productivity of Arabica as well as to control the White Stem Borer, the 12th Five Year Plan approved the Integrated Coffee Development Project (ICDP) subsidy scheme of the Coffee Board. Under the component of 'Development of Support for Traditional Areas', subsidies are provided to growers for gap filling in estates which are affected by the removal of White Stem Borer-infested Arabica plants. The support was provided to all small, medium and large growers, including corporate estates as well as co-operatives, subject to the condition that the estates have a minimum of 25 per cent vacancies. The subsidy was calculated taking into consideration the existing space in that area. The programme aimed to raise 50 lakh seedlings during 2015–16 for filling gaps in WSB-affected Arabica estates.

Besides the Mission Mode for tackling WSB, demonstrations for popularizing Integrated Pest Management (IPM), collaborative research with the Indian Council of Agricultural Research (ICAR) and the Bio Control Research Laboratories (BCRL) were the other measures initiated. Further, the Central Coffee Research Institute (CCRI) conducted field trials in 2014 to come up with solution to prevent the spread of WSB. The other measures include injections to the stem, coating with sealer-cum-healer, wrapping of the stem and feeding the roots with systemic pesticides.

5.7 Summary

During the survey, it was found that WSB was the major pest affecting Arabica plantations in Kodagu. About 68.33 per cent of the Arabica growers were shifting towards Robusta, while 21.67 per cent were willing to shift but unable to and 10 per cent were not willing to shift towards Robusta. Among the growers shifting towards Robusta, about 90 per cent strongly agreed that WSB was the major cause for the shift. The factors responsible for the WSB attack were classified based on the environmental and management factors. Among the environmental factors, the growers' ranked the rise in temperature as the number one cause of borer attacks, and among the management factors, they ranked the flight of WSB from the neglected neighbouring estates as the number one cause of borer attacks. The surveyed estates were classified according to elevation and rainfall as (LE + LR), (LE + MR), (LE + HR), (HE + LR), (HE + MR) and (HE + HR). The incidences of WSB were high in estates located at (LE + LR) estates, of which 20 per cent of estates suffered very high incidences, wherein uprooting ranged from 125 to 250 plants per ha.

Regression analysis has shown that among several factors, elevation had a significant influence (with a coefficient of −8.36) on the uprooting of arabica plants. Likewise, shade trees and pesticides also effectively controlled WSB. Our studies reveal that the incidence of WSB was higher in the plants coming under the age group of 20–30 years, while the uprooting of trees was lower in Arabica trees above 30 years. However, the area under Arabica cultivation and rainfall turned out to be insignificant. The cost incurred included chemical and cultural costs. The cultural costs varied between ₹670 and ₹7500 per ha, while the chemical costs varied between ₹7425 and ₹9600 per ha. The institutional support included the distribution of pheromone traps at subsidised rates, the pests and diseases act of 1968, the ICDP project, mission mode for WSB and collaborative research on WSB.

References

Ambinakudige, S.S., 2006, *Differential Impacts of Commoditization of Agriculture on people's Livelihoods and the environment in the Western Ghats of India: An extended environmental entitlement analysis.* Doctoral dissertation, University of Florida, Florida.

Basavaraj, M.C., Sreedharan, K., Vinod, K.P., Murphy, S.T. and Naidu, R., 2005, *Socio-economic analysis of smallholder coffee farms with reference to integrated management of white stem borer in Karnataka, India.* Proceedings of the ASIC 2004 20th International Conference on Coffee Science, Bangalore, 11–15 October 2004, pp. 1294–1298.

Belachew, M., 2003. *Coffee.* In Uhlig, S. (ed) *Encyclopaedia Aethiopica* 1. Harrassowitz, Wiesbaden, 763 pp.

Businessline, 2015, http://www.thehindubusinessline.com/economy/agri-business/research-body-develops-package-to-control-white-stem-borer-in-karnatakas-coffee-plantation

Chengappa, P.G., Rich, K.M., Muniyappa, A., Yadava, C.G., Ganashruthi, M.K., Pradeepa Babu, B.N., Shubha, Y.C. and Rich, M., 2014a. *Sustainability coffee certification in India.* Perceptions and Practices. NUPI Working Paper 830, Norwegian Institute of International Affair.

Davis, A.P., Gole, T.W., Baena, S. and Moat, J., 2012, The impact of climate change on Indigenous Arabica Coffee (Coffea arabica): Predicting future trends and identifying priorities. *PLoS One,* 7(11): e47981.

Deepika, M.G. and Jyotishi, A., 2013, *Assessing risk and risk mitigation strategies of small coffee growers: A study of Kodagu district in Karnataka,* NRPPD Discussion Paper No.21, Centre For Development Studies, Thiruvanathapuram.

Duffy, E.A.J., 1968, *A monograph of the immature stages of oriental timber beetles (Cerambycidae).* British museum (Natural History), London, pp. 202–207.

Garcia, C.A., Bhagwat, S.A., Ghazoul, J., Nath, C.D., Konerira, M.N., Cheppudira, K.G., Raghuramulu, Y., Nasi, R. and Vaast, P., 2010, Biodiversity conservation in agricultural landscapes: Challenges and opportunities of coffee agroforests in the Western Ghats, India. *Conservation Biology,* **24**(2): 479–488.

Hall, D.R., Cork, A., Phythian, S.J., Chittamuru S., Jayarama, B.K., Venkatesha, M.G., Sreedharan, K., Vinod Kumar, P.K., Seetharama, H.G. and Naidu, R., 2006, Identification of components of male-produced pheromone of coffee white Stemborer, Xylotrechus quadripes. *Journal of Chemical Ecology,* **32**: 195–219.

Indian Coffee Board. 2017, Database on Coffee, March 2017. Available at www.indiacoffee.org

Jayarama and Vinod Kumar P.K., 2009, Sex pheromones in the control of coffee white stemborer (*Xylotrechus quadripes*). *Indian Coffee,* **73**: 8–9.

Kumar, P.K.V., 2009, *Studies on the bioecology and interventions for the management of the coffee white stem-borer in India.* Project report CFC-ICO-18: Integrated Stem Borer Management in Smallholder coffee farms in India, Malawai and Zimbabwe.

Leroy, M., Garci, C., Aubert, P., Jeremy, V., Bernard, C., Brams, J., Caron, C., Junker, C., Payet., G, Rigal, C. and Thevenet, S., 2011, *Thinking the future: Coffee, forests and people – Conservation and development in Kodagu* Agro. ParisTech – ENGREF, Paris.

Murphy, T.S., Phiri, N.A., Sreedharan, K., Kutywayo, D. and Chanika, C., 2008, *Integrated stem borer management in smallholder coffee farms in India, Malawi and Zimbabwe.* Final Technical Report. CFC/CABI/ICO/CRS/CB.

Narayana, M.R., 2013, Does leaf rust matter for India's coffee farming? Evidence and implications from a recent household sample survey. *Journal of Plantation Crops*, **41**: 62–70.

Plat, J., 2012, Climate change threatens your morning cup of coffee. Available at http://climate-change-threatens-your-morning-cup-of-coffee

Radhakrishnan, S., Ramaiah, P.K. and Bhat, P.K., 1987, Methodology to estimate yield loss in coffee due insect pests. *Journal of Coffee Research,* **17**: 90–93.

Shylesha, A.N. and Veeresh, G.K., 1995, Incidence of coffee white stemborer *Xylotrechus quadripes* in major coffee growing tracts of Karnataka. *Journal of Hill Research*, **8**: 239–241.

Subramaniam, T.V. 1934. The coffee stemborer. *Entomological Series-Bulletin, Department of Agriculture (Mysore, India),* **11**: 1–18.

Titus, A. and Pereira, G.N., 2011, The fine art of irrigation in Robusta coffee Plantations. Available at https://ecofriendlycoffee.org/the-fine-art-of-irrigation-in-robusta-coffee-plantations

Venkatesha, M.G. and Dinesh, A.S., 2012. The coffee White Stem Borer *Xylotrechus quadripes* (Coleoptera: Cerambycidae): Bioecology, status and management. *International Journal of Tropical Insect Science*, **32**(4): 177–188.

Chapter 6

Economics of Arabica and Robusta coffee cultivation in Kodagu

6.1 Introduction

In the previous chapters, we have discussed the productivity of coffee and several factors influencing that productivity. However, higher productivity does not necessarily translate into high returns and high profitability, as the price and cost of cultivation are other variants that determine gross returns and net returns for a particular crop. Therefore, this section presents the economics of Arabica and Robusta varieties of coffee.

It is appropriate to understand the cost and returns on coffee to know the income-earning capacity of the growers. Besides, the cost of cultivation of coffee varies according to the varieties and sizes of holdings, which also has an impact on net returns. This information would serve the policy planners and researchers, who are the prime users of cost of cultivation for policy formulation at the microlevel. Apart from this, it would be appropriate to mention that, due to changes taking place both at the global and local levels, the importance of the cost of cultivation has increased manifold, besides being subject to constant changes. In this changing scenario, a comparison of the cost of production across different-sized holdings and varieties is a practical necessity. Hence, an attempt has been made to understand cost variations and returns across various sizes of holdings.

The components provided under the 'comprehensive scheme for studying the cost of cultivation of principal crops' have been re-arranged and adapted according to varieties and land holdings of coffee to shed light on the fixed and variable costs of Arabica and Robusta varieties. Further, the total revenue generated has been estimated, as the costs and returns aspects play a vital role in effective cost control, profit planning and sales decisions of the growers.

DOI: 10.4324/9781003541721-6

6.2 Total operational costs (₹/ha) incurred in Arabica and Robusta varieties of coffee

Tables 6.1a and 6.2a present the costs and net returns per hectare realized by different-sized growers during 2015–16 for Arabica and Robusta varieties of coffee. The cost incurred was disaggregated into two components, viz. operational cost and fixed cost.

The operational costs are the costs that are directly associated with the cultivation and harvesting of coffee crops. The total operational costs accounted for 62 per cent and 64 per cent of the total costs in Arabica and Robusta, respectively, while the fixed cost was 38 per cent and 36 per cent in Arabica and Robusta, respectively. The total operational costs in the case of Arabica increased with the size of holdings, viz. ₹92,232, ₹100,356, ₹107,964 and ₹112,847 per ha among the small, semi-medium, medium and large growers, respectively. The percentage of operational costs to the total cost decreased with the size of holdings, which was found to be 65 per cent, 62 per cent, 61 per cent and 59 per cent in the case of small, semi-medium, medium and large growers, respectively (Table 6.1a). In the case of Robusta, the total operational costs were ₹83,943 per ha for small holdings, ₹91,712 per ha for semi-medium holdings, ₹93,984 per ha for medium holdings and ₹105,602 per ha for large growers, on an average constituting 64 per cent of the total cost of cultivation.

The predominant cost was the labour cost that comprised 55 per cent of the total operational costs. The labour costs comprised wages paid for hired labour and the imputed value of family labour, which accounted for 58 and 42 per cent of the total labour cost, respectively, in Arabica. In the case of Robusta, the hired labour and family labour constituted 56.91 and 43.09 per cent of the total labour cost, respectively. The labour costs in Arabica were found to be highest among the large growers (₹59,228 per ha), followed by semi-medium (₹57,124 per ha), medium (₹56,987 per ha) and small growers (₹53,109 per ha) (Table 6.1a). In the case of Robusta, the labour cost was found to be lowest among the medium growers (₹49,476 per ha), followed by semi-medium (₹50,102 per ha), small (₹50,118 per ha) and large growers (₹56,859 per ha). The overall labour cost in Robusta was ₹50,655 per ha (Table 6.2a).

In the case of small growers, generally, family labour was used for plant maintenance, being assisted by hired labour, while, the hired labour increased with the increase in size of holdings in both the varieties. It was found to be highest among the large growers (₹54,175 per ha in Arabica and ₹53,147 per ha in Robusta). The hired labour cost was lowest among the small growers (₹10,037 per ha in Arabica and ₹14,506 per ha in Robusta). Coffee cultivation is a labour-intensive and time-sensitive process that cannot be easily mechanized. Therefore, the hiring of non-family labour was an indispensable part of coffee cultivation in Kodagu.

The family labourers were highest in terms of imputed costs among the small growers (₹43,072 and ₹35,612 per ha in Arabica and Robusta, respectively), semi-medium (₹29,527 and ₹26,533 per ha in Arabica and Robusta, respectively), medium (₹9582 and ₹8364 in Arabica and Robusta, respectively) and large growers (₹5053 and ₹3712 per ha in Arabica and Robusta, respectively). The overall cost of family labour was found to be ₹23,748 and ₹21,828 per ha in Arabica and Robusta, respectively. In the case of Arabica, the per cent cost of family labourers to total labour cost was found to be highest among small growers (81.10 per cent), followed by semi-medium (51.69 per cent), medium (16.81 per cent) and large growers (8.53 per cent) (Table 6.1a). Likewise, in Robusta, the per cent of family labour to total labour costs decreased with the size of holdings, viz. small (71.06 per cent), semi-medium (52.96 per cent), medium (16.91 per cent) and large (6.53 per cent) (Table 6.2a).

The input costs ranged between 18.73 and 19.81 per cent of the total operational costs in Arabica, whereas the range was found between 29 and 33 per cent in Robusta. Among the input costs, the fertilizers accounted for 60 per cent and 66 per cent of the input costs in Arabica and Robusta, respectively. The overall cost of fertilizers was found to be ₹19,065 per ha and ₹19,569 per ha in Arabica and Robusta, respectively. In the case of Arabica, the fertilizer cost was found to be highest among the large growers (₹20,492 per ha), followed by medium (₹20,180 per ha), semi-medium (₹18,583 per ha) and small growers (₹17,694 per ha) (Table 6.1a). In the case of Robusta, the fertilizer cost was highest among large growers (₹21,250 per ha) followed by semi-medium growers (₹19,909 per ha), medium (₹19,551 per ha) and small growers (₹16,889 per ha).

The pesticides in Arabica constituted about 16.87 and 4.82 per cent of the total input costs in Arabica and Robusta, respectively, accounting for ₹5870 per ha in large growers, followed by medium (₹5649 per ha), small (₹5695 per ha) and semi-medium growers (₹5019 per ha) (Table 6.1a). In the case of Robusta, the pesticide cost was found to be ₹1171 per ha among the large growers, followed by semi-medium (₹1126 per ha), medium (₹1109 per ha) and small growers (₹1088 per ha). The overall costs of pesticides in Arabica, viz. Chloropyrifos, Lindane, Contof, lime and neem accounted for 16.87 per cent of the total input cost, whereas in the case of Robusta it was relatively lower, at 3.80 per cent of the total input cost (Table 6.2a).

The overall irrigation costs for Arabica and Robusta were ₹5528 per ha and ₹6610 per ha, respectively. The highest irrigation cost was found to be among the large growers (₹6998 per ha in Arabica and ₹7592 per ha in Robusta), followed by medium (₹5742 per ha and ₹7103 per ha in Arabica and Robusta, respectively), semi-medium (₹5401 per ha in Arabica and ₹6484 per ha in Robusta) and small (₹4201 and ₹5675 per ha in Arabica and Robusta, respectively). In the case of Arabica, the irrigation costs constituted 19 per cent of total input costs among the large growers, semi-medium

(17.74 per cent), medium (16.85 per cent) and small growers (14.90 per cent) (Table 6.1). However, in the case of Robusta, the irrigation costs comprised 23.06 per cent of total input costs in small, semi-medium (21.32 per cent), medium (23.19 per cent) and large growers (23.44 per cent).

The liming and dolomite were practically used to overcome the acidic soils to restore the normal pH of the soil. While the liming and dolomite constituted about 1.21 per cent and 1.42 per cent of the total input costs in Arabica and Robusta, respectively, the application of liming and dolomite was not found among the small Arabica growers. The overall costs for these items were found to be ₹385 per ha and ₹425 per ha in Arabica and Robusta, respectively. The highest cost in Arabica was observed among the medium growers (₹568 per ha), followed by large (₹542 per ha) and semi-medium (₹367 per ha) growers (Table 6.1a). In Robusta, the highest cost was observed among the large growers (₹489 per ha), followed by semi-medium (₹441 per ha), medium (₹431 per ha) and small (₹296 per ha) growers (Table 6.2a).

The weeding constituted 4.46 and 7.13 per cent of the total input costs in Arabica and Robusta, respectively. In the case of Arabica, it was found to be highest among the large growers (₹2749 per ha), whereas in the case of Robusta, it was highest among medium growers (₹2425 per ha). The overall cost of weeding was higher in Robusta, viz. ₹2131 per ha, as compared to Arabica, ₹1418 per ha.

The transplanting, gap filling and fence keeping were included under the maintenance costs. In Arabica, transplanting and gap filling costs were found to be highest among the medium growers (₹1999 per ha), followed by small growers (₹1694 per ha), large growers (₹1378 per ha) and semi-medium growers (₹1325 per ha) (Table 6.1a). Whereas transplanting costs were relatively lower among the Robusta growers, it was found to be higher among the large growers (₹166 per ha), medium (₹146 per ha), small (₹139 per ha) and semi-medium (₹72 per ha) (Table 6.2a). The transplanting and gap filling accounted for about 86.67 per cent and 28.07 per cent of the total maintenance costs in Arabica and Robusta, respectively. The perentage of transplanting as part of total maintenance cost in Arabica was highest among small growers (92.87 per cent), followed by large growers (89.13 per cent), semi-medium (88.03 per cent) and medium growers (80.93 per cent). Meanwhile, the cost of fencing accounted for 13.32 per cent and 71.92 per cent of the total maintenance costs in Arabica and Robusta, respectively (Table 6.1a and 6.2a)..

The processing costs accounted for about 66.67 per cent and 61.46 per cent of total processing costs in Arabica and Robusta, respectively. The processing cost was highest among medium growers (₹1983 per ha), followed by large (₹1923 per ha), semi-medium (₹1781 per ha) and small (₹1755 per ha) growers in Arabica. In the case of Robusta, the highest processing costs were observed among the medium growers (₹935 per ha), followed by large (₹904 per ha), semi-medium (₹842 per ha) and small (₹834 per ha). The costs of processing materials (water, machinery rent and electricity costs) accounted for 33.32 per cent of total processing costs in Arabica (Table 6.1a). The processing cost was found to be highest among the medium (₹994 per ha), followed by large (₹965 per ha), semi-medium (₹889 per ha) and small growers (₹877 per ha). In Robusta, the costs of materials accounted for about 38.54 per cent of the total processing costs. In Robusta, the material cost was highest among semi-medium growers (₹592 per ha), followed by medium (₹583 per ha), large (₹490 per ha) and small (₹314 per ha).

The marketing costs included the transportation, storage, labour charges, and the value of bags or baskets. The value of gunny bags accounted for about 62.94 per cent and 70 per cent of the marketing costs in Arabica and Robusta, respectively, followed by the transportation charges (23 per cent), storage charges (6.68 per cent), and labour charges for loading and unloading (7.30 per cent). The overall marketing cost was found to be highest among the large growers (₹1852 per ha), followed by medium (₹1707 per ha), semi-medium (₹1598 per ha), and small growers (₹1605 per ha) in Arabica (Table 6.1a). In Robusta, the marketing cost was highest among large growers (₹4359 per ha), followed by medium (₹3451 per ha), semi-medium (₹2985 per ha) and small (₹2914 per ha). The overall marketing cost was found to be ₹3205 per ha in Robusta (Table 6.2a).

6.3 Total fixed costs incurred in Arabica and Robusta varieties

The rent on leased land or the imputed value of rent on owned land, land revenue, cesses and taxes, depreciation on implements and buildings, interest on fixed capital and establishment costs were taken into account for estimating the fixed costs. The total fixed costs were found to be highest among the large growers (₹78,863 per ha and ₹67,510 per ha in Arabica and Robusta, respectively), followed by medium (₹68,646 per ha and ₹56,197 per ha in Arabica and Robusta, respectively), semi-medium (₹62,085 per ha and ₹49,716 per ha in Arabica and Robusta, respectively) and small growers (₹50,049 per ha and ₹45,939 per ha in Arabica and Robusta, respectively). The overall fixed costs were 38 per cent and 36.10 per cent of the total costs in Arabica and Robusta, respectively. The fixed costs of the large growers accounted for 41 per cent of the total costs, followed by medium (39 per cent), semi-medium (38 per cent), and small growers (35 per cent) in Arabica. In the case of Robusta, the per cent of fixed to total costs was highest among the large growers (39 per cent), followed by medium (37 per cent), semi-medium (35 per cent) and small (35 per cent) (Tables 6.1a and 6.2a).

The imputed rent on owned land comprised 42.69 per cent of the total fixed cost in Arabica, which was found to be highest among small growers (52.74 per cent), followed by semi-medium (44.54 per cent), medium (39.69

per cent) and large growers (34.88 per cent). In the case of Robusta, the imputed rent on owned land comprised 58.80 per cent of the total fixed cost, and it was lowest among the large (44.45 per cent), followed by medium (54.91 per cent), semi-medium (62.50 per cent) and small (64.74 per cent) growers. The overall rent on owned land was ₹27,382 per ha in Arabica and ₹30,745 per ha in Robusta, with the highest among semi-medium growers (₹27,658 per ha and ₹31,071 per ha in Arabica and Robusta, respectively), followed by large growers (₹27,514 per ha and ₹30,011 per ha in Arabica and Robusta, respectively), small growers (₹26,396 per ha and ₹29,742 per ha in Arabica and Robusta, respectively) and medium growers (₹27,251 per ha and ₹30,860 per ha). The land revenue, cesses, and taxes comprised 0.05 per cent and 0.07 per cent of total fixed costs in Arabica and Robusta, respectively.

The amortized establishment costs as a per cent of total fixed costs were found to be 29.46 per cent and 17.65 per cent in Arabica and Robusta, respectively. In Arabica, the percentage was highest among small growers (33.13 per cent), followed by semi-medium (30.53 per cent), medium (29.55 per cent) and large (23.74 per cent) growers (Table 6.1a). In the case of Robusta, the percentage was highest among semi-medium growers (18.67 per cent), followed by small growers (17.22 per cent), medium growers (17.16 per cent) and large growers (14.62 per cent) (Table 6.2a). The overall establishment cost was ₹18,894 and ₹9229 per ha in Arabica and Robusta, respectively. In Arabica, this cost was highest among medium growers (₹20,288 per ha), followed by semi-medium (₹18,952), large (₹18722 per ha) and small (₹16,579 per ha) growers, while the amortized establishment cost in Robusta was highest among large growers (₹9867 per ha), followed by medium growers (₹9645 per ha), semi-medium growers (₹9281 per ha) and small farmers (₹7911 per ha).

The overall depreciation of farm implements and buildings was found to be ₹10,969 per ha and ₹6681 per ha in Arabica and Robusta, respectively. In both varieties, the highest cost was observed among the large growers (₹24,157 and ₹20,370 per ha in Arabica and Robusta, respectively), followed by medium growers (₹13,733 per ha and ₹9642 per ha in Arabica and Robusta, respectively), semi-medium growers (₹8802 per ha and ₹4008 per ha in Arabica and Robusta, respectively) and small growers (₹1689 per ha and ₹3333 per ha in Arabica and Robusta, respectively). The depreciation constituted 17.10 and 12.78 per cent of the total fixed costs, which was found to be highest among the large growers (30.63 and 30.17 per cent in Arabica and Robusta, respectively), followed by medium (20.01 per cent and 17.16 per cent in Arabica and Robusta, respectively), semi-medium (14.18 per cent in Arabica and 8.06 per cent in Robusta) and small (3.37 per cent in Arabica and 7.26 per cent in Robusta). Meanwhile, the interest on fixed capital was found to be ₹6855 per ha in Arabica (10.69 per cent of the total fixed costs) and ₹5594 per ha in Robusta (10.70 per cent of the total fixed costs) (Tables 6.1a and 6.2a).

6.4 Total costs and revenue in coffee cultivation

The total cost in Arabica was found to be ₹142,280, ₹162,441, ₹176,609 and ₹191,710 per ha in small, semi-medium, medium and large growers, respectively. The overall total cost in Arabica was ₹166,697 per ha. In Robusta, the total cost was ₹129,881, ₹141,428, ₹150,181 and ₹173,111 per ha in small, semi-medium, medium and large growers, respectively.

The output per ha increased with the size of holdings. The overall output was found to be 1004 kg parchment per ha and 3180 kg of cherry per ha in Arabica and Robusta, respectively. In Arabica, it was found to be 969 kg/ha among small, 1010 kg/ha among semi-medium, 1000 kg/ha among medium and 1023 kg/ha among large growers (Table 6.1a). However, in Robusta, it was 3150 kg/ha among small, 3176 kg/ha among semi-medium, 3258 kg/ha among medium and 3175 kg/ha among large growers.

The overall price per kg of Arabica parchment was observed to be ₹164 per kg, while that for Robusta cherry was ₹58 per kg. The total revenue in Arabica increased with the size of holdings, viz. small (₹158,129 per ha), semi-medium (₹163,128 per ha), medium (₹163,500 per ha) and large (₹168,334 per ha). However, in Robusta, it was found to be ₹179,550 for small, ₹181,159 per ha for semi-medium, ₹192,222 per ha for medium and ₹185,198 per ha for large growers.

However, the total cost was higher than the total revenue among medium and large Arabica growers, who incurred losses of ₹13,109 per ha and ₹23,376 per ha, respectively. In contrast, semi-medium and small growers earned profits of ₹687 per ha and ₹15,849 per ha, respectively. However, in the case of Robusta, growers earned an overall profit of ₹39,852 per ha. The profit was found to be ₹49,669 per ha for small, ₹39,731 per ha for semi-medium, ₹42,041 per ha for medium and ₹12,086 per ha for large growers.

6.5 Different components of cost of production in Arabica and Robusta coffee

The cost of production (₹/ha) in Arabica and Robusta varieties was estimated separately across the size of holdings. Cost A_1 included all actual expenses in cash and kind incurred by the growers plus amortized establishment cost and depreciation of fixed assets, which was found to be ₹108,712 per ha and ₹86,640 per ha in Arabica and Robusta, respectively. The Cost A_1 for small growers was ₹67,466 per ha for Arabica and ₹59,612 per ha for Robusta; for semi-medium growers, it was ₹98,621 and ₹78,506 per ha for Arabica and Robusta, respectively; for medium growers, it was ₹132,440 and ₹104,945 per ha for Arabica and Robusta, respectively; and for large growers, it was ₹150,711 and 132,164 per ha for Arabica and Robusta, respectively (Table 6.3a).

Cost B_1 includes Cost A_1 and interest on the value of owned fixed capital other than the land. The overall Cost B_1 values were found to be ₹115,567 and ₹92,234 per ha in Arabica and Robusta, respectively. The Cost B_1 values for Arabica small, semi-medium, medium and large growers were estimated to be ₹72,813, ₹105,256, ₹139,776 and ₹159143 per ha, respectively. The Cost B_1 for Robusta small growers was ₹64,527 per ha; semi-medium growers, ₹83,824 per ha; medium growers, ₹110,957 per ha; and large growers, ₹139,388 per ha (Table 6.3a).

Cost B_2 included Cost B_1 and the rental value of owned land. The values of Cost B_2 for small, semi-medium, medium and large growers in Arabica were found to be ₹99,209, ₹132,914, ₹167,027 and ₹186,657 per ha, respectively. The overall Cost B_2 for Arabica was ₹142,949 per ha. In Robusta, the overall Cost B_2 was ₹122,979 per ha. The Cost B_2 for Robusta small was observed as ₹94,269 per ha; semi-medium, ₹114,895 per ha; medium, ₹141,817 per ha; and large growers, ₹169,399 per ha (Table 6.3a).

Cost C_1 comprised Cost B_1 and the imputed value of family labour. The Cost C_1 for small growers was ₹115,885 per ha and ₹100,139 per ha for Arabica and Robusta, respectively; semi-medium growers, ₹134,783 and 110,357 per ha for Arabica and Robusta, respectively; medium growers, ₹149,358 per ha and ₹119,321 per ha for Arabica and Robusta, respectively; while for large growers, it was ₹164,196 for Arabica and ₹143,100 for Robusta. The overall cost of cultivation was ₹139,315 and ₹114,062 per ha for Arabica and Robusta, respectively.

Cost C_2 comprises Cost B_2 and imputed value of family labour. The overall values of Cost C_2 were found to be ₹166,697 per ha and ₹144,810 per ha for Arabica and Robusta, respectively. The values of C_2 were estimated to be ₹142,281 and ₹129,881, respectively, for Arabica and Robusta; for semi-medium, ₹162,441 and ₹141,428 per ha for Arabica and Robusta, respectively; for medium, ₹176,609 and ₹150,181 per ha for Arabica and Robusta, respectively; and for large, ₹191,710 and ₹173,111 per ha for Arabica and Robusta, respectively (Table 6.3a).

As compared to the cost of cultivation, there was a significant variation in the cost of production (₹/qtl.) between Arabica and Robusta varieties due to differences in the processing of these two types. Owing to the high cost of cultivation, the overall cost of production in the Arabica variety was higher. Hence, the Cost A_1 for Arabica worked out to be ₹10,828 per quintal, while for Robusta it worked out to be ₹2725 per quintal. The Cost A_1 values for small growers were ₹6964 and ₹1892 per quintal for Arabica and Robusta, respectively; for semi-medium, ₹9764 and ₹2472 per quintal for Arabica and Robusta, respectively; for medium, ₹13,244 per quintal for Arabica and ₹3221 per quintal for Robusta; and for large growers, ₹14,726 and ₹4163 per quintal, respectively, for Arabica and Robusta (Table 6.3a).

The overall values of Cost B_1 in Arabica and Robusta were ₹11,511 and ₹2900 per quintal, respectively. They were found to be ₹7516, ₹10,421, ₹13,978 and ₹15,550 per quintal for small, semi-medium, medium and large Arabica growers, respectively. In contrast, it worked out to be ₹2048, ₹2639,

₹3406 and ₹4390 per quintal for small, semi-medium, medium and large Robusta growers, respectively.

Cost C_2 takes into account all other non-paid and imputed cost items like family labour, land rent and interest on fixed assets, which worked out to be ₹16,603 and ₹4554 per quintal in Arabica and Robusta, respectively. The Cost C_2 values for small, semi-medium, medium and large growers were ₹14,687, ₹16,082, ₹17,661 and ₹18,732 per quintal, respectively. In the case of Robusta, the Cost C_2 values were found to be ₹4123, ₹4453, ₹4610 and ₹5452 per quintal, respectively (Table 6.3a).

6.6 Returns from coffee cultivation

The overall net income over total cost from coffee cultivation was ₹15,848 per ha among the small Arabica growers, followed by semi-medium (₹687 per ha), while it was found to be negative among the medium (–₹13,109) and large growers (–₹23,376). In the case of Robusta, the net income was higher: small growers (₹49,669 per ha), semi-medium (₹39,731 per ha), medium (₹40,868 per ha) and large (₹12,086 per ha). Overall, the Arabica growers were incurring losses from coffee cultivation, while Robusta growers were earning reasonably good returns of ₹39,852 per ha.

As seen from Table 6.4a, net income was estimated by deducting Cost C_2 from gross income, viz. net income over total cost. The net income over total cost turned out to be negative in the case of medium and large Arabica growers, viz. –₹13,109 and –₹23,376, respectively, which was mainly attributed to higher fixed costs among the medium (₹68,646 per ha) and large (₹78,863 per ha) growers, which were higher than the average fixed costs of the Arabica growers (₹64,138 per ha). The higher fixed costs were also due to allowing non-cash items such as depreciation and amortized establishment costs. These non-cash expenses even made considerable variation across the size of holdings, highest among the large growers wherein the total costs exceeded the total revenue. For instance, depreciation accounted for merely 3 per cent of total fixed costs among small Arabica growers, whereas it accounted for 31 per cent of total fixed costs among the large Arabica growers. The depreciation taken together with amortized establishment costs accounted for 37 per cent of total fixed costs, whereas in the case of medium and large growers, it accounted for 54 per cent and 57 per cent of the total fixed costs, respectively. Non-cash items like depreciation and amortization as part of fixed costs have reduced the net income of the medium and large growers significantly. These fixed costs along with the variable costs were reflected in terms of higher total costs. Henceforth, the higher total costs for medium and large growers when deducted from the gross income led to negative net income.

Among the fixed costs, there was a substantial difference in the costs of depreciation across the size of the holdings, viz. ₹1689, ₹8802, ₹13,733 and

₹24,157 per ha for small, semi-medium, medium and large growers, respectively, as the value of machineries or implements was found to be higher among large Arabica growers (Table 3.4a). Besides, depreciation cost was higher among large growers, as larger machines were expensive and tended to depreciate faster. Furthermore, the amortized establishment cost was higher among large growers. Both depreciation and amortized establishment costs are considered non-cash items.

The returns over variable costs varied between ₹90,663 and ₹17,623 per ha in Arabica, with the highest returns among semi-medium growers (₹64,507 per ha), followed by small growers (₹90,663 per ha), large growers (₹17,623 per ha) and medium growers (₹31,060 per ha). However, in Robusta, the returns over variable costs were found to be highest among semi-medium growers (₹102,653 per ha), followed by medium (₹86,104 per ha), large (₹53,033 per ha) and small (₹119,938 per ha). The overall returns over variable costs amounted to ₹98,023 per ha. The farm business income, which represents returns over Cost A_2, was similar to the returns over variable costs as there was no difference between Cost A_1 and Cost A_2 as the leasing of land was not found among the surveyed growers (Table 6.4a).

The returns over variable costs are calculated as gross income less Cost A_1, an indicator of a farm's ability to meet financial obligations. Therefore, in this study, returns over variable costs are used to estimate the relative profitability of different varieties of coffee across land holdings. The returns over variable costs were found to be lower among Arabica growers (1.76 times). This was mainly attributed to higher input and labour costs in Arabica as compared to Robusta. Besides, the overall Cost A1 in Arabica was found to be ₹108,712 per ha as against Robusta (₹86,640 per ha). Even the gross income of Arabica growers was found to be lower, viz. ₹164,335 per ha as compared to ₹184,663 per ha for Robusta. Henceforth, wide variations were observed in returns over variable costs.

The family labour income per hectare of Arabica cultivation at an overall level was found to be ₹21,386 per ha, with small growers earning ₹58,920 per ha and semi-medium, ₹30,214 per ha and medium and large growers incurring losses of –₹3527 and –₹18,323 per ha, respectively. In Robusta, it was found to be ₹85,281 per ha among small, ₹49,232 per ha among medium, ₹66,264 per ha among semi-medium and ₹15,798 per ha among large growers (Table 6.4a).

The overall returns per rupee of investment were found to be ₹0.99 in Arabica and ₹1.28 per ha in Robusta. In the case of Arabica, returns were highest among small growers, viz. ₹1.11, followed by semi-medium (₹1.00), medium (₹0.93) and large (₹0.88). In the case of Robusta, it was found to be ₹1.38 among small, followed by semi-medium (₹1.28), medium (₹1.27) and large (₹1.07). The returns per rupee of investment in Robusta were higher (₹1.28) than in Arabica (0.99). In other words, returns per rupee of investment in Robusta were 1.99 times higher than in Arabica. Therefore, the hypothesis that the Robusta variety offers higher returns on investment than the Arabica variety in the study region can be accepted. Additionally, it can be observed from the table that the returns per rupee of investment were higher among small growers, followed by semi-medium, medium and large growers, which were 1.11, 1.00, 0.93 and 0.88, respectively, in Arabica and 1.38, 1.28, 1.27 and 1.07 in Robusta, respectively. Thus, there existed an inverse relation between farm size and rate of returns. The higher returns per rupee of investment reflect an economically efficient coffee production enterprise and vice versa, more prominently observed among the small and medium Arabica growers.

6.7 Gross income vs total resources used in production

Having expended many of the resources in coffee cultivation, it becomes very essential to determine the revenue or income accrued per quantity of resource used in production. Thus, it is imperative to understand the contribution of each resource to the gross income accrued by the grower. This section attempts to determine the total revenue per unit of resource used by Arabica and Robusta growers in Kodagu. Therefore, the gross income (₹/ha) of coffee growers was regressed on labour use (man-days), fertilizers (kg/ha), irrigation (₹/ha), pesticides (barrels/ha) and weedicides (litres/ha). The functional form of the model is given below.

$$\ln(\text{Gross income}) = \alpha_0 + \beta_1 \ln(\text{Labour use}) + \beta_2 \ln(\text{fertilizers}) + \beta_3 \ln(\text{irrigation}) + \beta_4 \ln(\text{pesticides}) + \beta_5 \ln(\text{weedicides}) + e$$

where Y = gross income per ha
α = constant or intercept
e = unexplained error term

Coffee production is a labour-intensive and an expensive process. Given the shortage of the labour force in the coffee sector, does an increase in quantity (or number) of labour actually increase the gross income of the coffee growers? In order to answer this question, the quantity of labour (man days/ha) was a major input included in the model. Secondly, fertilizer was also a major input for coffee that accounted for 60 per cent and 66 per cent of the total input costs in Arabica and Robusta, respectively. Our survey revealed that growers applied variable amounts of fertilizers, and the frequency of fertilizer application (twice or thrice) also varied between the growers. Given the variation in fertilizer quantity, the study attempts to analyze its impact on gross income in both varieties of coffee.

Generally, Robusta growers rely heavily on sprinkler irrigation. On the other hand, Arabica growers spend less on sprinkler irrigation given the rain-fed nature of Arabica, besides limited access to irrigation in these regions. Considering irrigation as an important factor that promotes uniform flowering and fruit setting in coffee, the cost of irrigation (₹/ha) was taken into account to understand its influence on the gross Income of the grower.

The coffee crop suffers heavy yield losses due to pests, insects and nematodes. Chemical pesticides were thereby applied in coffee production to protect the crop from pests such as insects, mites, nematodes, ticks and rodents, which could cause losses to the crop. The use of chemical pesticides (insecticides, nematicides and fungicides) has increased in recent years for effective pest control and also to increase the output. Therefore, whether chemicals actually enhance the profits of coffee was the key question. Thus, the model takes into account the quantity of pesticide use (barrels/ha) to understand its influence on the gross Income of the growers.

Similarly, the presence of weeds within the coffee plantations adversely affects the output. While the application of weedicides such as Glycel or Roundup varied across the varieties, does the quantity of weedicides (litres/ha) on coffee plantations have any impact on the gross income of the coffee growers? In order to understand this aspect, weedicide (litres/ha) was included in the model.

Labour use (man-days) was significant in both cases, indicating that every 1 per cent increase in the number of labour man-days increased the gross income (GI) by 0.28 per cent in Arabica and 0.18 per cent in Robusta, both being significant at the 1 per cent level of significance. The coefficient associated with fertilizer also turned positive and significant. In the case of Arabica, it was found that every 1 per cent increase in fertilizer increased the gross income by 0.39 per cent. In the case of Robusta, every 1 per cent increase in fertilizers increased the gross income by 0.15 per cent. The irrigation costs in Arabica increased the gross income by 0.01 per cent, whereas in Robusta they increased the output by 0.21 per cent. The quantity of pesticides, though insignificant in Robusta, however, increased the gross income of Arabica growers by 0.23 per cent. Our study reveals that weedicides had no impact on the gross income of coffee as the coefficient associated with weedicides was insignificant in both varieties. Hence, from the above results, it can be concluded that inputs or resources had a major influence on the gross Income of the growers (Table 6.5a).

6.8 Summary

Total costs per ha incurred by growers increased with an increase in size of holdings in both varieties. The cost of cultivation of Arabica was ₹166,697 per ha, of which 62 per cent was operational costs and 38 per cent was fixed costs. In Robusta, the total cost of cultivation was ₹144,811 per ha of which operational costs accounted for 64 per cent of the total costs and fixed costs accounted for 36 per cent of the total costs. Labour costs accounted for 34 per cent and 35 per cent of total costs in Arabica and Robusta, respectively. The returns per rupee were highest among the small Robusta growers, followed by semi-medium growers. In Arabica, returns per rupee were highest among small growers. With these costs, at the prevailing prices, medium and large growers incurred losses of ₹13,379 and ₹26,426 per ha, respectively.

The returns over variable costs were found to be higher in Robusta growers (₹98,023 per ha) compared to Arabica (₹55623 per ha) growers. The family labour income was found to be negative among medium (–₹3527) and large (–₹18,323) Arabica growers. Similarly, the net income over total cost was also found to be negative among medium and large Arabica growers, viz. –₹13,109 and –₹23,376, respectively. Meanwhile, the returns per rupee of investment were higher among Robusta growers (1.28) compared to Arabica growers (0.99).

Labour use (man-days), fertilizers (kg/ha) and irrigation application (₹/ha) had a positive and significant influence on the gross income of the growers in both varieties,while pesticide application had a positive influence on Arabica, weedicides had no impact on the gross income of the growers in both varieties. Thus, it can be concluded that resource use in coffee production had a positive and significant influence on the gross income of growers.

Chapter 7

The market structure and pattern of coffee in Kodagu

7.1 Introduction

The market structure refers to the formal organization of the functional activity of a marketing institution that influences the nature of competition and pricing policy in the market. In other words, market structure is a combination of factors that determines the interaction between buyers and sellers and price changes in the market, while the market pattern is the sequence through which the transfer of ownership occurs from the point of production to consumption.

The schematic representation of the current coffee marketing structure in Kodagu is shown in Figure 7.1. At each stage of marketing, a certain amount of value addition takes place. The value chain is the structure of activities that focuses on creating value for the product at each stage of marketing. Accordingly, Chengappa et al. (2014a) mapped the value chain for conventional, certified and organic coffee in Kodagu, as shown in Figures 7.2, 7.3 and 7.4, respectively.

Figure 7.2 shows the value chain structure for conventional coffee that comprises several actors, viz. coffee growers, village-level consolidators/collectors, local traders, hullers, companies, importers, roasters and consumers. At the producers' level, primary processing occurs wherein wet or dry processing takes place. Generally, the Robusta growers produce dry cherries, while the Arabica growers wet process their produce to form parchments. The small growers who lack infrastructure for the preparation of parchment would hire the processing machines. The coffee is then marketed either to local traders, hullers, curers, companies or exporters. The secondary processing is carried out at the curing works or by companies like NED Commodities – a Netherlands-based export house, Ecom Gill and SLN – a Swiss-based export house and Ned Commodities who in turn market them to domestic roasters, wholesalers or retailers or export the produce.

The value chain structure for certified coffee is presented in Figure 7.3. Coffee certification is a process wherein independent third-party certification agencies provide written assurance that the product complies with certain

DOI: 10.4324/9781003541721-7

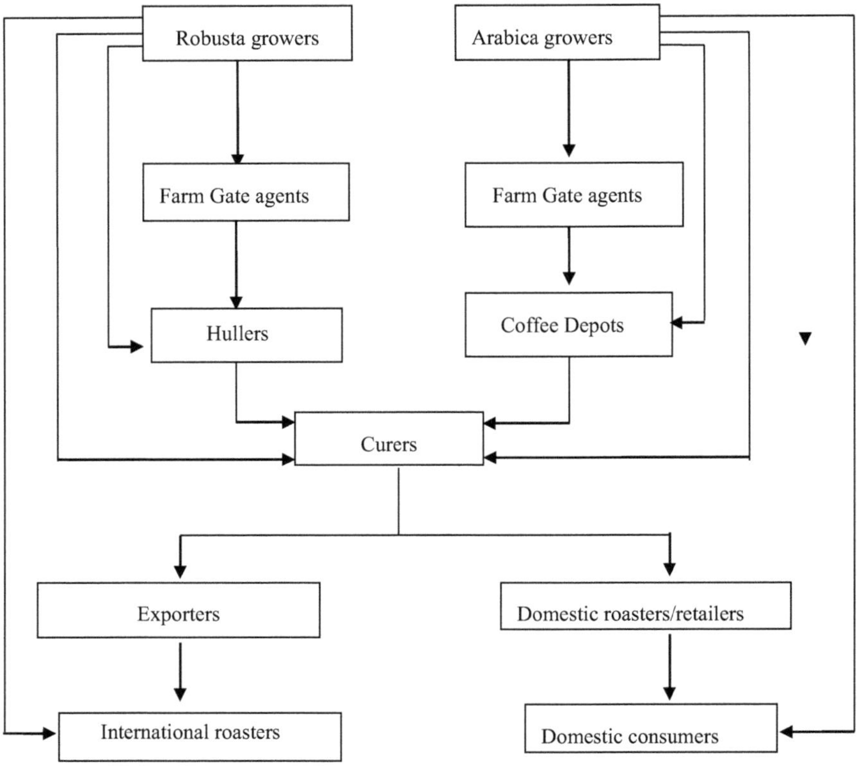

Figure 7.1 The general market structure of coffee in Kodagu.

Source: Field survey.

social, economic and environmental norms. Generally, it takes about three to six months for growers to prepare the plot for Utz and Rainforest Alliance (RFA) certification. In Kodagu district, Ecom-Gill (a Swiss-based export house) and Ned Commodities (a Netherlands-based export house) form a group of small growers for certification. This process reduces the costs of certification while covering a larger area under group certification. The cost of certification for Rainforest Alliance (RFA) is approximately ₹250 and ₹375 per ha for small and large growers, respectively. In the case of Utz, the cost of certification worked out to be ₹1000 per ha, along with separate fees for annual audits and travel charges. In the case of group certification, the certificate is issued in the name of companies or export houses. To encourage the growers to undertake certification, companies also bear the cost of certification for the first two or three years and support the producers in realizing higher premiums for their produce, besides providing them with opportunities to access new markets.

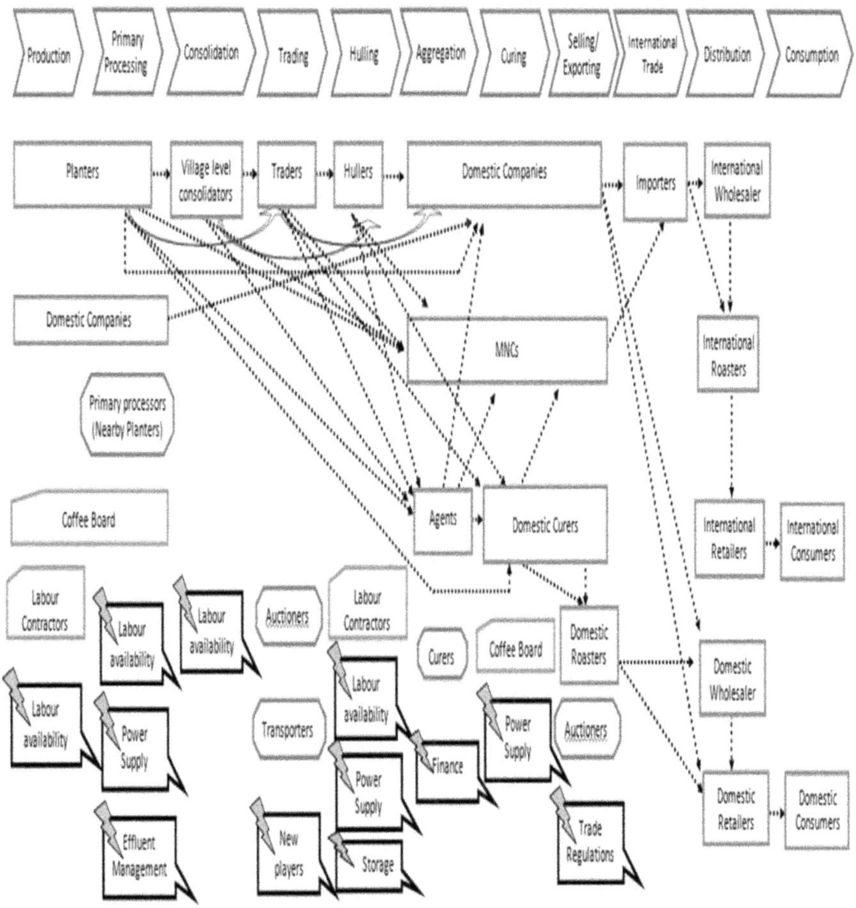

Figure 7.2 Value chain structure for conventional coffee.

Source: Chengappa et al., 2014b.

The value chain structure for organic coffee is provided in Figure 7.4. For the grower to get his farm certified as organic, it requires about three years for conversion, known as 'conversion period'. In this case, the Coffee Board has endorsed three certification agencies, namely IMO (Institute for Market-ecology), Skal International and APOF (Association for Promotion of Organic Farming), for certification of organic coffee in India in accordance with National Programme for Organic Production (NPOP) norms. Once the coffee grower is in compliance with these norms, a certificate is issued and labelled as 'certified organic'. However, the major threats to organic production remain the difficulty in obtaining organic inputs, initial yield reduction,

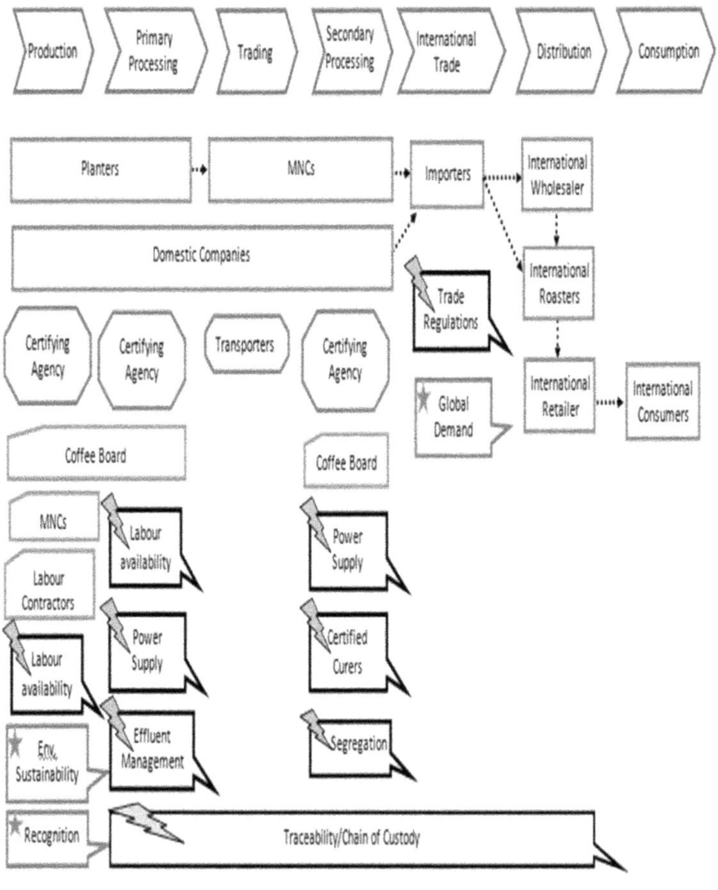

Figure: 7.3 Value chain structure for certified coffee.

Source: Chengappa et al., 2014b.

higher costs of certification and lower price premiums relative to the initial costs.

7.2 Mode of sales of coffee

In general, the mode of sales undertaken by the coffee producers in the study area is provided in Table 7.1a. The prices of farm gate sales over the years are provided in Figure 7.5. There are several reasons for such farm gate sales. Firstly, the immediate requirement for cash needs, especially among the small and semi-medium growers, wherein the revenue received from such sales immediately contributes to the growers' income and meets the domestic

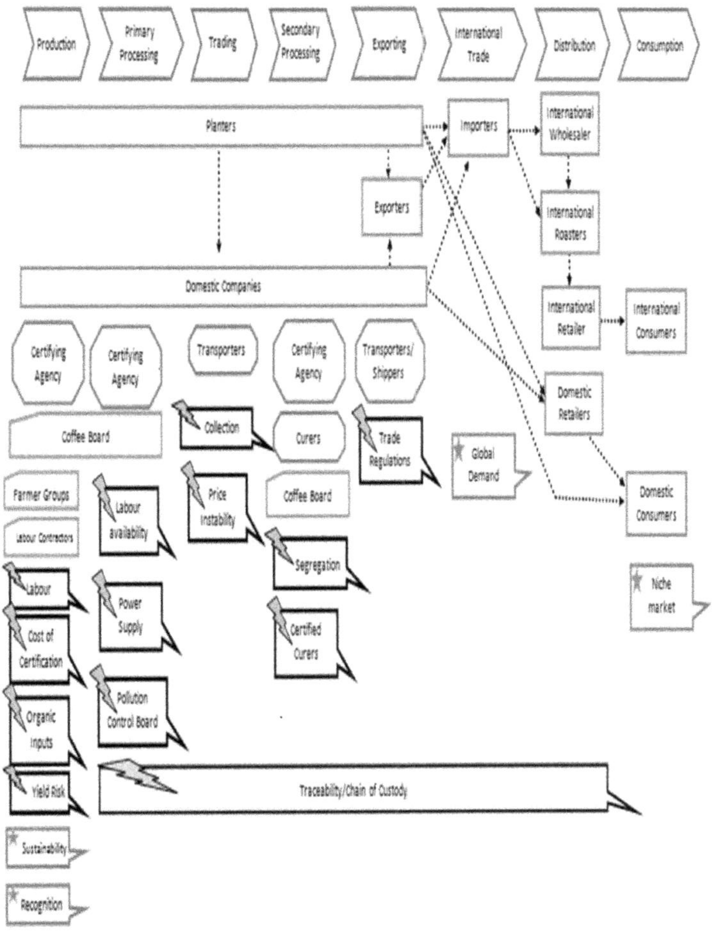

Figure 7.4 Value chain structure for organic coffee.

Source: Chengappa et al., 2014b.

needs. The growers also make informal contracts with private traders who satisfy the financial needs of the growers well in advance of the harvesting season, which also force them to sell their produce at the farm gate itself.

Secondly, Rule 7B of Income Tax Rules[1] 1962, wherein the growers are taxed for value addition, has further forced the growers to sell their coffee at the farm gate itself without any processing or value addition. Further, the decision of farm gate sales is also influenced by inadequate transport facilities, poor infrastructure and risks involved in arranging such transport. However, the local traders at the farm gate arrange for their own transport,

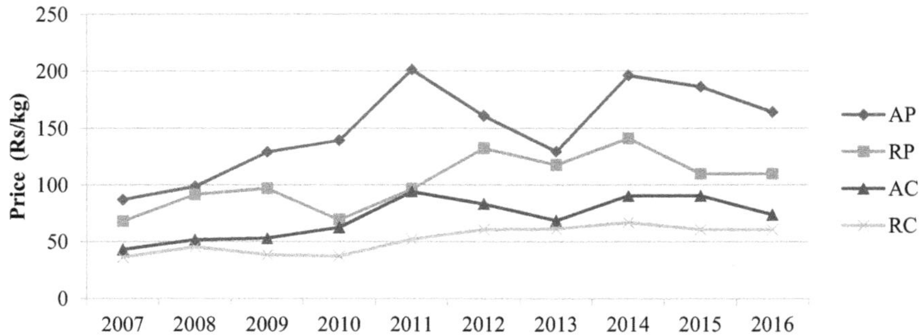

Figure 7.5 Farm gate prices for different varieties of coffee (₹/kg). AP – Arabica Parchment; AC – Arabica cherry; RP – Robusta Parchment; RC – Robusta Cherry.

gunny bags and loading of coffee onto the trucks, thereby relieving growers of transportation and other risks.

7.2.1 Fluctuations in monthly farm gate prices in Kodagu

At a microlevel in the district, in the case of Arabica Parchment, the average deviation of the monthly farm gate prices and coefficient of variations over a decade was found to be highest during 2014 (Table 7.2a), while price fluctuations were lowest during 2007, wherein the intra-year coefficient was also the lowest during the period. The highest price for Arabica Parchment generally occurred in the first quarter (January to March) during the years 2007, 2012, 2013 and 2015, while the same quarter also experienced the lowest prices during the years 2008, 2009 and 2011. The lowest prices were also witnessed during the fourth quarter of 2012, 2013 and 2015. Thus, it can be concluded that the first quarter witnessed the highest price variation over the past few years, followed by the fourth quarter. Interestingly, these two quarters also coincide with the onset of the procurement season for Arabica. For instance, in our study, about 81.66 per cent of the growers marketed coffee during this period, which were subject to maximum price fluctuations and price risks. Generally, the growers sell the produce during this period owing to immediate financial needs required for harvesting and post-harvest operations.

On the other hand, in the case of Robusta Cherry, the average deviation in monthly farm gate prices was found to be highest during 2012. The coefficient of variation was found to be 10.07 per cent, while the standard deviation was lowest during 2009, with the coefficient of variation during the year being 4.13 per cent (Table 7.3a). The highest prices for Robusta Cherry over the years were found in the fourth quarter (October to December), which were evident during 2007, 2010, 2011 and 2016. On the other hand, the lowest farm gate prices for Robusta Cherry were found during the first quarter between January and March (2007, 2008, 2011 and 2012). In the case of Robusta, generally, the harvest season begins in the month of December, and growers were not able to take advantage of higher prices during this season. The peak Robusta sales commenced during the first quarter of the year, which also coincided with the lowest prices for Robusta, as shown in Table 7.3a. About 83 per cent of Robusta growers marketed a part of their produce during this quarter owing to financial needs for immediate irrigation requirements for Robusta for blossoming. In this case, also, maximum price variations were observed during the month of January, while the highest prices were evident during 2009, 2013 and 2015. The lowest prices during January were observed in the years 2008 and 2011.

The value chain structure for certified coffee is presented in Figure 7.3. Coffee certification is a process wherein the independent third-party certification agencies provide a written assurance of the product that complies with certain social, economic and environmental norms. Generally, it takes about three to six months for growers to prepare the plot for Utz and Rainforest Alliance (RFA) certification, while in Kodagu district, Ecom-Gill (a Swiss-based export house) and Ned Commodities (a Netherlands-based export house) form a group of small growers for certification. This process reduces the costs of certification while covering a larger area under group certification. Approximately the cost of certification for Rainforest Alliance (RFA) worked out to be ₹250 and ₹375 per hectare for small and large growers, respectively, whereas, in the case of Utz, the cost of certification worked out ₹1000 per hectare, and there exists separate fees for annual audits and travel charges. In the case of group certification, the certificate is issued in the name of companies or export houses. To encourage the growers to undertake certification, the companies also bear the cost of certification for the first two or three years and support the producers in realizing higher premiums for their produce, besides providing them with opportunities to access new markets.

Generally, the coffee stocks start depleting in the month of October/November, while the local actors (especially the roasters) look forward to starting to rebuild their depleted stocks, which usually takes place in the month of January. Therefore, much demand is created in the month of January as the local markets start purchasing the produce at the beginning of the harvest (January), which is reflected in the form of wider price fluctuations during this month.

The monthly price variations of Arabica Cherry in Madikeri over the years are shown in Table 7.4a. The highest standard deviation was observed in 2012, wherein the co-efficient of variation was 10.07 per cent. The lowest standard deviation was observed during the year 2009. In this case, the highest farm gate prices were observed in the fourth quarter between October and December (2007, 2009, 2010, 2011 and 2016). The highest farm gate prices for Arabica Cherry were observed during the month of December (Table 7.4a).

Rather than farm gate sales, about 30 per cent of Arabica growers sold their produce to coffee depots. There were a series of coffee collection centres in the Arabica growing belts of Kodagu (particularly Somwarpet). These depots acted as mere coffee collecting centres and neither undertook the processing of coffee, nor were responsible for transportation costs, which were to be borne by the growers. These collection centres also collected small quantities of Liberica variety of coffee (*Coffea liberica*) at ₹30 per kg and transported them to Kerala for medicinal purposes. They purchased coffee at the prevailing market price and in turn sold the produce to the exporters (NED Commodities, Ecom Gill and SLN) with whom they had entered into earlier contracts.

About 23 per cent of surveyed Arabica and 18 per cent of the Robusta growers sold their coffee to the curing works or processing units, where secondary processing of coffee was undertaken. At the curing works, the coffee was mainly tested for moisture, out-turn per cent, Berry Borer, blacks, browns and bits. Here, the peeling machines remove the outer husk and segregate the beans into different grades. In the large curing works, the defective beans were usually handpicked by skilled women, where each and every bean was examined before packing to maintain a very high quality.

While there are several curing works located in the coffee-growing regions in Karnataka, Kerala and Tamil Nadu (Figure 7.6), our study region comprised 16 licensed curing works, which acted as chief nodes in the marketing of coffee, usually owned by the exporters. The curing works also provided for consignment sales wherein the growers stored the coffee at the warehouses of curing works that retained ownership of the produce and sold the coffee during times of higher prices, thereby taking advantage of higher price movements. The curing works were also required to maintain the quality of coffee as per the requirements of Indian Coffee Board standards (Table 7.5a).

About 15 per cent of Robusta growers sold their produce to the hullers, who acted as intermediaries between growers and exporters, a practice absent among the Arabica growers. The milling process (husk or parchment

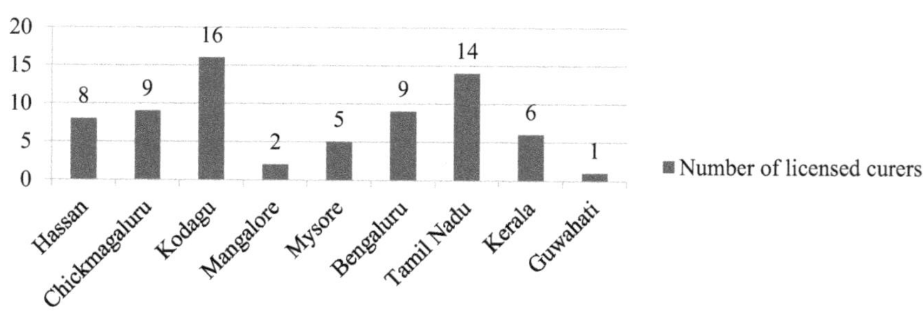

Figure 7.6 Number of licenced coffee curing works in India (2015–16).

Source: Indian Coffee Board, 2017.

removal) was done, and the product was converted to green beans. The hullers owned machinery for milling of coffee, while local traders rented the hullers' pulpers and used their drying yards for drying coffee. Small traders also sold their cherries to the hullers. Nevertheless, the industry opinion is that the deterioration of quality at this stage generally occurs, which is primarily attributed to the poor processing practices followed by the hullers, and this was responsible for a higher quantity of damaged beans in the coffee lots.

On the other hand, about 12 per cent each of Arabica and Robusta growers sold their produce to the companies (SLN Coffee Private Limited, Tata Coffee Ltd, Nestle and Ecom Gill Coffee Trading Private Limited), who purchased both certified and non-certified coffee. About 2 per cent of Arabica and 3 per cent of Robusta growers directly exported their produce, while only 2 per cent of Arabica growers were involved in online sales of coffee.

Nevertheless, about 23.33 per cent and 25 per cent of surveyed Arabica and Robusta growers were certified. The Utz certification was commonly observed among the growers. in both Arabica and Robusta varieties, due to the group certification initiatives. This was followed by Rainforest Alliance (RFA) and Nespresso certification among the Arabica growers, while the Robusta growers followed Common Codes for Coffee Community (4C) and Organic certification (Table 7.6a). However, none of the small growers (<2 ha) adopted certification programmes due to barriers such as costs, knowledge and tedious record-keeping process.

7.2.2 *Why do growers end up selling certified coffee in the mainstream market?*

In our survey, it was revealed that only 50 per cent of certified Arabica and 46.67 per cent of certified Robusta growers actually sold their produce as certified, while the rest sold the produce as non-certified coffee (Table 7.7a). Even at the global level, the actual sales of certified coffee represented only a minor portion of production. Over 2008–2012 it was estimated that only one-fourth of the sustainably produced coffee was 'actually sold' as certified in the market (Table 7.7a), thereby receiving extra benefits, higher premiums and preferential market access. The remaining coffee, though certified for sustainable practices, was merely sold as 'ordinary' coffee in the conventional market without any market benefits. The major reasons quoted by the growers were that the assured price premiums were much lower than the existing market prices. Moreover, if the existing certified stocks were not sold during a specific period, the validity of such sustainable certification would expire, forcing them to sell it in the conventional market. Even at the global level the sales of certified coffee were less than half the production (Table 7.8a).

7.3 Market integration in coffee

Market Integration refers to the expansion of firms by consolidating additional market functions and activities under a single management (Kohls et al., 1980). Market integration also refers to the interrelationship between price movements in two markets. The measurements of the extent of market integration are attempted at two different levels: Integration among firms within a market and integration among spatially separated markets. In this section, we attempt to measure integration in spatially separated markets through price cointegration.

Figure 7.7 shows the movement of Farm Gate prices of Arabica Parchment, Arabica Cherry and Robusta Cherry with international prices taken from January 2000 to November 2017. The descriptive statistics of Farm Gate prices of Arabica Parchment, Arabica Cherry, Robusta Cherry and international prices of Arabica and Robusta are provided in Table 7.9a. In the case of international Arabica price, the maximum price (₹16,541 per 50 kg) was observed during April 2011, whereas the maximum international price for Robusta (₹6723 per 50 kg) was found during May 2011.

The average farm gate prices for Arabica Parchment, Arabica Cherry and Robusta Cherry were found to be ₹5671, ₹2744 and ₹2000 per 50 kg, respectively. However, the average international prices for Arabica and Robusta were found to be ₹7531 and ₹4030 per 50 kilograms, respectively. The minimum price of ₹1800 for Arabica Parchment was observed in February 2002, while the minimum prices of Arabica Cherry and Robusta Cherry were ₹788 and ₹444 per 50 kilograms, respectively, observed in January 2002 and February 2002. The Arabica international price (₹2991 per 50 kg) was

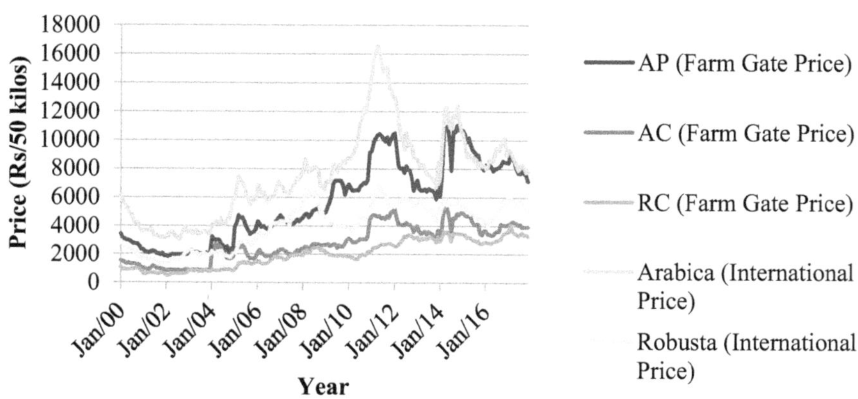

Figure 7.7 Movement of Farm Gate Prices (₹/50 kilo of Medikeri - Kadagu district) with international prices over the years (2000-2017). AC-Arabica Cherry; AP–Arabica Parchment; and RC-Robusta Cherry.

lowest in August 2002, while the Robusta international price was ₹1257 per 50 kilograms in January 2002. The possible cause of the lower price was overproduction of coffee relative to its actual demand. Conversely, the maximum farm gate prices for AP, AC and RC per bag were found to be ₹11,037 (October 2014), ₹5313 (March 2014) and ₹4075 (February 2017). The maximum Arabica International prices were ₹16,541 per bag (April 2011), while the maximum Robusta International prices were ₹6723 per bag (May 2011).

7.3.1 Unit root test

The 'augmented Dickey–Fuller' (ADF) test was used to test whether time series data were stationary or not. This was used for determining the order of integration of time series, which requires regressing ΔP_{it} on a constant, P_{it-1} and several lags of ΔP_{it}.

$$\Delta P_{it} = \alpha + \beta P_{it-1} + \sum_{k=1}^{M} \delta_k \Delta P_{it-k} + e_{it}.$$

Then, the '*t*' statistic of the estimated coefficient of P_{it--1} can be used to test the null and alternative hypothesis viz., H_c: $P_{it} \sim I\ (1)$ vs. H_1: $P_{it} \sim I\ (0)$.

The critical values larger than the standard critical *t* values and negative are called 'ADF Statistics', rather than the *t* statistics. In case the null hypothesis that has been set out cannot be rejected, the time series P_{it} is termed as non-stationary with either $I\ (1)$ or $I\ (2)$, else even have a higher order of integration. To understand the order of integration, the test could be repeated with ΔP_{it} as an alternative for P_{it}. Therefore, regressing $\Delta^2 P_{it}$ on a constant, ΔP_{it-1} and on various lags of $\Delta^2 P_{it}$ viz.,

$$\Delta^2 P_{it} = \alpha + \beta \Delta P_{it-1} + \sum_{k=1}^{M} \delta_k \Delta^2 P_{it-k} + e_{it}.$$

Here, the following tests are conducted by the ADF statistics:

$$H_0: \Delta P_{it} \sim I\ (1) \text{ vs } H_1: \Delta P_{it} \sim I\ (0)$$

$$\text{i.e., } H_0: P_{it} \sim I\ (2) \text{ vs } H1: P_{it} \sim I\ (1).$$

The null hypothesis (H_0) cannot be rejected if the ADF statistic is not large and negative and that P_{it} cannot be I (1). Here, in this case, there is repetition with $\Delta^3 P_{it}$ as the dependent variable, and so forth until the determination of the order of integration is decided. In order to avoid auto-correlated disturbances, the lagged terms in ΔP_{it} are included. Excluding this, the t statistic on P_{it-1} is considered as ordinary and not the augmented Dickey–Fuller test or as DF test.

Now, taking the regression equation

$P_{it} = \rho P_{it-1} + e_t$, where e_t is the classical error term. The general assumption is that in the absence of loss of generosity that $\rho > 0$, known as the Autoregressive AR (1) Model, and is stationary, if and only if $|\rho| < 1$ and because

$$V(P_{it}) = \frac{\delta^2}{(1-\rho)^2}$$

This should be non-negative and then the only likelihood is that $|\rho| = 1$. P_{it} would be a random walk and has a unit root, when at the extreme $\rho = 1$. Thus, the Dickey–Fuller test is also known as a unit root test. After deducting P_{it-1} from both sides and rearranging the same, the equation is written as follows, known as Dickey–Fuller regression, with the assumption that

$\rho > 0$.

$$\Delta P_{it} = (\rho - 1)P_{it-1} + e_t$$

Now, the null hypothesis $P_{it} \sim I$ (1) is equivalent to the coefficient on P_{it-1} in the Dickey–Fuller regression being zero. Then for the I (0) process, one can expect the one-sided that the coefficient on $P_{it-1} < 0$ for P_{it} to be stationary.

After identifying the order of integration of individual prices, the co-integration between them needs to be identified. When the two series are co-integrated, they are tied together, even though they are trending. While the theory of co-integration establishes that the even though the two time series (P_{it} and P_{jt}) are integrated, u_t non-stationary, some other linear combinations in them appears to be stationary. Therefore, price integration would mean that a long-run relationship exists between the two price series after the transient effects have disappeared.

In order to test whether the two series are co-integrated, the 'two-step' procedure by Engel–Granger is followed, wherein the first step involves the co-integration of P_{it} and $P_{jt,}$ those which have the same order of integration, i.e. I (1). Here, the co-integrating vector is stationary after first differencing and has the same order of integration, I (1).

Prior to the implementation of the efficiency test, it is necessary to assess the level of integration in the coffee markets. There is substantial evidence in the literature that the unit root tests of the Dickey–Fuller perform better when the lag length (p) is chosen from the data according to a model selection or information criterion, like the Schwarz information criterion, than when the lag length p has been set to an arbitrary large number. Here, Dickey–Fuller unit root tests were used to verify that Arabica and Robusta prices were non-stationary.

As shown in Table 7.1, the Augmented Dickey–Fuller (ADF) test shows that for all five series, the level of data was non-stationary. Here, the ADF test does not reject the unit test root hypothesis. However, stationarity was reached after the first difference. This implies that the data was integrated of order one, $I(1)$. In case of the first difference of the series, the hypothesis was, however, rejected, mostly at the 1 per cent level, indicating that all five series of coffee were stationary of order one.

7.3.2. Johansen's co-integration method

Post unit root testing and with the assumption that the time series data are integrated of the same order, the Johansen test is conducted to investigate co-integration involving all variables. This is a maximum likelihood method that determines the number of co-integrating vectors in a non-stationary time series. The Vector Auto Regression (VAR) with restrictions imposed is also known as the Vector Error Correction Model (VEC). Johansen's estimation model is given as follows:

$$\Delta X_t = \mu + \sum_{i=1}^{p} r_i \Delta X_{it-1} + \alpha\beta X_{t-I} + \varepsilon_t,$$

$X_t = (n \times 1)$ vector of all the non-stationary indices

$r_i = (n \times n)$ matrix of coefficients

$\alpha = (n \times r)$ matrix of error correction coefficients where r is the number of co-integrating relationships in the variables, such that $0 < r < n$, which measures the speed at which the variables adjust their equilibrium, also known as the adjustment parameter.

$\beta = (n \times r)$ matrix of r co-integrating vectors, so that $0 < r < n$,

which represents the long-run co-integrating relationship between the variables. Using Akaike's (AIC) and Schwarz's Bayesian (SBIC) Information Criterion process, the lag length for Johansen's procedure was determined.

Johansen uses two different test statistics, viz. the Trace Test method and the maximum Eigen value test for co-integrating. The Trace Test is a joint test for testing the null hypothesis of no co-integration (H_0: $r = 0$) against the alternative hypothesis of co-integration (H_1: $r \geq 0$). In contrast, the maximum

Table 7.1 Augmented Dickey–Fuller (ADF) test for unit root: Arabica and Robusta prices[2]

Augmented Dickey-Fuller (ADF) Test of unit root for domestic and international markets

Series	*Sample: 2000 (Jan) to 2017 (Nov)*							
	Levels				*First Difference*			
	t-stat	*Prob*	F *Test*	*Remarks*	*t-stat*	*Prob*	F *Test*	*Remarks*
Log Arabica Cherry	−1.43	0.56	3.48	Non-stationary	−7.33	0	24.04	Stationary
Log Arabica International	−1.27	0.64	5.71	Non-stationary	−11.78	0	138.88	Stationary
Log Arabica Parchment	−1.44	0.55	2.56	Non-stationary	−3.7	0	21.89	Stationary
Log Robusta Cherry	−0.69	0.84	0.48	Non-stationary	−13.34	0	178.13	Stationary
Log Robusta International	−1.07	0.72	2.93	Non-stationary	−12.53	0	157.15	Stationary

Source: Own compilation.

Eigen value test carries out test on each of the eigen values separately. Here, the null hypothesis is tested for the number of co-integrating vectors that is equal to r against the alternative of $r + 1$ co-integrating vectors.

$$\lambda_{trace}(r) = -T \sum_{i=r+1}^{g} \ln(1 - \hat{\lambda}_i)$$

$$\lambda_{max}(r, r+1) = -T \ln(1 - \hat{\lambda}_{r+1})$$

r = number of co-integrating vectors under null
$\hat{\lambda}_i$ = estimated ith order eigen value from the $\alpha\beta'$ matrix

Here, a significantly non-zero eigen value represents a significant co-integrating vector.

The results of Johansen's Trace and max Eigen value tests are presented in Table 7.2 It indicates the non-existence of co-integration between Arabica Cherry and International markets. Similar results were observed for Robusta Cherry and Robusta international, as shown in Table 7.3, for Arabica Parchment and Arabica International (Table 7.4), Robusta International and Arabica International (Table 7.5), Arabica Cherry, Arabica Parchment and Arabica International (Table 7.6).

Following the unit root, the next step used was the Johansen procedure for testing the co-integration of domestic and international coffee prices, to verify if there exists any long-run relation between these prices. The results of Johansen's Trace and Max Eigen value tests are presented in Table 7.8. They indicate the existence of one co-integrating equation at the 1 per cent level of significance for Arabica Cherry markets. For Arabica Cherry markets, the maximum Eigen value test also shows one co-integrating equation at the 1 per cent level, confirming the Trace Test results. Both the maximum Eigen value test and the Trace Test results for Arabica Cherry markets indicate the presence of one co-integrating vector at 1 per cent level of significance. This implies that the Arabica Cherry market was integrated and had a common sharing of information on price changes in the long run.

7.3.3 Granger causality test

Granger (1969) proposed an approach in order to determine the causality, an approach to examine whether x causes y; to understand to what extent the present value of y could be explained by the past values of y; and further

Table 7.2. Johansen's maximum Eigen value and Trace Test for the number of co-integrating vectors – Arabica Cherry and Arabica international

Johansen's Maximum Eigen value and Trace Test for the number of co-integrating vectors

Sample: Jan 2000 to Nov 2017

Series tested for co-integration	*Number of co-integrating vectors*	*Eigen value*	*Trace statistic*	*Probability*	*Maximum-Eigen*	*Probability*	*Order of VAR*
Log Arabica Cherry	0	0.037	9.46	0.32	8.11	0.36	1 to 4
Log Arabica International	1	0.0064	1.35	0.24	1.35	0.24	

Source: Own compilation.

Table 7.3 Johansen's Maximum Eigen value and Trace Test for the number of co-integrating vectors – Robusta Cherry and Robusta International

Johansen's Maximum Eigenvalue and Trace Test for the number of co-integrating vectors

Sample: 2000 (Jan) to 2017 (Nov)

Series tested for co-integration	*Number of co-integrating vectors*	*Eigen value*	*Trace statistic*	*Probability*	*Maximum Eigen*	*Probability*	*Order of VAR*
Log Robusta cherry	0	0.039	9.40	0.32	8.48	0.33	1 to 4
Log Robusta International	1	0.004	0.92	0.33	0.92	0.33	

Source: Own compilation.

Table 7.4 Johansen's Maximum Eigen value and Trace Test for the number of co-integrating vectors – Arabica Parchment and Arabica international

Johansen's Maximum Eigen value and Trace Test for the number of co-integrating vectors

Sample: 2000 (Jan) to 2017 (Nov)

Series tested for co-integration	*Number of co-integrating vectors*	*Eigen value*	*Trace statistic*	*Probability*	*Maximum Eigen*	*Probability*	*Order of VAR*
Log Arabica Parchment	0	0.03	9.46	0.32	8.11	0.36	1 to 4
Log Arabica International	1	0.03	1.35	0.24	1.35	0.24	

Source: Own compilation.

Table 7.5 Johansen's Maximum Eigen value and Trace Test for the number of co-integrating vectors – Robusta International and Arabica International

Johansen's Maximum Eigenvalue and Trace Test for the number of co-integrating vectors							
Sample: 2000 (Jan) to 2017 (Nov)							
Series tested for co-integration	*Number of co-integrating vectors*	*Eigen value*	*Trace statistic*	*Probability*	*Maximum Eigen*	*Probability*	*Order of VAR*
Log Robusta International	0	0.02	7.34	0.53	6.03	0.60	1 to 4
Log Arabica International	1	0.006	1.31	0.25	1.31	0.25	

Source: Own compilation.

Table 7.6 Johansen's Maximum Eigen value and Trace Test for the number of co-integrating vectors – Arabica Cherry, Arabica Parchment and Arabica international

Johansen's Maximum Eigen value and Trace Test for the number of co-integrating vectors

Sample: 2000 (Jan) to 2017 (Nov)

	Number of co-integrating vectors	*Eigen value*	*Trace statistics*	*Probability*	*Maximum Eigen*	*Probability*	*Order of VAR*
Log Arabica Cherry	0	0.066	23.57	0.21	14.38	0.33	1 to 4
Log Arabica Parchment	1	0.036	9.19	0.34	7.83	0.39	
Log Arabica International	2	0.006	1.35	0.24	1.35	0.24	

Source: Own compilation.

Table 7.7 Johansen's Maximum Eigen value and Trace Test for the number of co-integrating vectors – Robusta Cherry, Robusta international, Arabica international and Arabica Cherry

Johansen's Maximum Eigen value and Trace Test for the number of co-integrating vectors

Sample: 2000 (Jan) to 2017 (Nov)

	Number of co-integrating vectors	*Eigen value*	*Trace*	*Probability*	*Maximum Eigen*	*Probability*	*Order of VAR*
Log Robusta Cherry	0	0.14	47.76**	0.05	32.08***	0.01	1 to 4
Log Robusta International	1	0.04	15.68	0.73	8.10	0.89	
Log Arabica International	2	0.03	7.57	0.51	6.75	0.51	
Log Arabica Cherry	3	0.003	0.81	0.36	0.81	0.36	

Source: Own compilation.
Note: *** and ** significant at 1 per cent and 5 per cent level of significance.

Table 7.8 Johansen's Maximum Eigen value and Trace Test for the number of co-integrating vectors

Johansen's Maximum Eigen value and Trace Test for the number of co-integrating vectors							
Sample: 2000 (Jan) to 2017 (Nov)							
Series tested for co-integration	*Number of co-integrating vectors*	*Eigen value*	*Trace statistic*	*Probability*	*Maximum-Eigen*	*Probability*	*Order of VAR*
Log Arabica Cherry	0	0.19	76.45***	0.01	44.72***	0.001	1 to 4
Log Arabica International	1	0.07	31.73	0.62	16.58	0.61	
Log Arabica Parchment	2	0.03	15.14	0.77	7.94	0.90	
Log Robusta Cherry	3	0.029	7.20	0.55	6.18	0.58	
Log Robusta International	4	0.005	1.02	0.31	1.02	0.31	

Source: Own compilation.
Note: *** denotes rejection of the hypothesis at 1 per cent significance level.

to examine if adding the lagged values of x could enhance the explanation. For instance, if y is Granger-caused by x, then x supports in the prediction of y, viz. x Granger causes y and y Granger causes x. In general, the model is given as

$$AI^3 i = \sum_{i=1}^{m} a_i AI_{t-i} + \sum_{i=1}^{m} b_i AC_{t-i} + V_t$$

$$AC^4_i = \sum_{i=1}^{m} a_i AC_{t-i} + \sum_{i=1}^{m} b_i AI_{t-i} + V_t$$

Since the series of prices are non-stationary and integrated of order one I (1), the first difference is taken in lieu of it being stationary (examined with the Augmented Dickey–Fuller Test). The minimization of Akaike's Final Prediction Error (FPE) helps to determine the optimal lag length. From the above equation, AC Granger causes AI if H_0: $b_1 = b_2 = b_3 = \ldots = b_n = 0$ is rejected against the alternative H_1: at least one $b_i \neq 0$, $i = 1,\ldots,n$.

Johansen's co-integration test performed above indicates only the existence of a long-run relationship of coffee prices. However, to determine the direction of the relationship, Granger causality tests were performed. As shown in Table 7.9, this test was based on 213 observations and was performed with a lag value of 4.

The results reveal that there exists bi-directional flow of price information from AI to AC and AC to AI, from AP to AI and AI to AP. Meanwhile, uni-directional causality is revealed from AI to RI and RI to RC. This uni-directional causality, wherein the prices of AI and RI markets lead to the prices in other markets, viz. RI and RC, without having any feedback impact on the prices, in turn, reveals that the market is not very efficient. The F-statistics for RI to RC is significant at the 1 per cent and AI to RI significant at 5 per cent level of significance. The bi-directional causation is observed in AI to AC and AC to AI, the former significant at the 1 per cent, while the latter significant at the 10 per cent level of significance. Another bi-directional flow is from AI to AP and AP to AI, the former significant at the 5 per cent and latter significant at the 1 per cent level of significance. Interestingly, the bi-directional causation is observed in Arabica markets, viz., Arabica

Table 7.9 Granger causality tests between Arabica and Robusta prices in domestic and international markets (N = 213)

Direction	*Null hypothesis*	F* *Statistics*	P-*value*	*Decision of null hypothesis*	*Remarks*
AI → AC	AI does not Granger cause AC	4.86***	0.0009	Rejected	Bi-directional
AC → AI	AC does not Granger cause AI	2.24*	0.06	Rejected	
AI → AP	AI does not Granger cause AP	9.33***	6.00E-07	Rejected	Bi-directional
AP → AI	AP does not Granger cause AI	2.24**	0.05	Rejected	
AP →AC	AP does not Granger cause AC	1.36	0.24	Accepted	No causality
AC →AP	AC does not Granger cause AP	1.40	0.23	Accepted	No causality
RI →AC	RI does not Granger Cause AC	2.11	0.18	Accepted	No causality
AC →RI	AC does not Granger cause RI	0.32	0.86	Accepted	No Causality
RC →AC	RC does not Granger cause AC	1.50	0.2	Accepted	No causality
AC →RC	AC does not Granger cause RC	0.79	0.53	Accepted	No causality
RI →AI	RI does not Granger cause AI	0.45	0.77	Accepted	No causality
AI →RI	AI does not Granger cause RI	2.65**	0.03	Rejected	Unidirectional
RC →AI	RC does not Granger cause AI	1.47	0.21	Accepted	No causality
AI →RC	AI does not Granger cause RC	3.70	0.6	Accepted	No causality
RI →AP	RI does not Granger cause AP	3.42	0.9	Accepted	No causality
AP → RI	AP does not Granger cause RI	0.84	0.5	Accepted	No causality
RC →AP	RC does not Granger cause AP	2.45	0.14	Accepted	No causality
AP →RC	AP does not Granger cause RC	2.04	0.19	Accepted	No causality
RC → RI	RC does not Granger cause RI	0.39	0.81	Accepted	No causality
RI → RC	RI does not Granger cause RC	8.25***	3.00E−06	Rejected	Unidirectional

Source: Own compilation.
Note: ***, ** and * significant at 1%, 5% and 10% respectively.

International, Arabica Cherry and Arabica International markets, wherein AI leads to AC and AP, AP and AC to AI, thereby implying the existence of efficient marketing.

The *p*-value of 0.009 means the null hypothesis that AI does not Granger cause AC can be rejected for the function. Likewise, in the direction of AC to AI, significant at 10 per cent that Arabica Cherry does not Granger cause Arabica International could also be rejected. In the case of AI to AP, the p-value being significant at 1 per cent, the hypothesis that Arabica International does not Granger cause Arabica Parchment can be rejected. Similarly, the hypothesis that Robusta International does not Granger cause Robusta Cherry can be rejected.

On the other hand, the null hypothesis that AP does not Granger cause AC, AC does not Granger cause AP, RI does not Granger cause AC, AC does not Granger cause RI, RC does not Granger cause AC, AC does not Granger cause RC, RI does not Granger cause AI, RC does not Granger cause AI, AP does not Granger cause RI and RC does not Granger cause RI could be accepted.

7.4 Price spread and marketing efficiency of coffee

From the earlier discussion, we saw very feeble integration between the domestic and international prices. Now, in this section, we proceed to understand the price spread in the domestic market by taking into account the price received by the growers at the farm gate and the net margin of several intermediaries involved in coffee marketing. This study helps us to determine the producers' share in the consumer's rupee and the price spread. Here, the price spread comprises of marketing costs and marketing margins that in turn determine the overall efficiency of the marketing system. Henceforth, the marketing costs, margins, price spread, and marketing efficiency have been worked out and are presented in Table 7.10.

The price per kg of Robusta sold by the roaster-cum-retailer was ₹300 per kg, while the producers received a net price of only ₹56.80, ₹57.60 and ₹60.20 per kg in channel 1, channel 2 and channel 3, respectively. The expenses incurred by producers varied between ₹1.40 and ₹1.80 per kg of Robusta. In the case of Arabica, the selling price of roaster-cum-retailers was found to be ₹450 per kg. The growers received only ₹165, ₹166.20 and ₹167.10 in channel 4, channel 5 and channel 6, respectively. The expenses incurred by the producers varied between ₹1.50 and ₹1.90 per kg of Arabica (Table 7.10).

The marketing efficiency of dried Robusta was calculated using Acharya's Modified Measure of Marketing Efficiency (MME). The marketing efficiency in channel 1 and channel 2 was 0.19, whereas it was marginally higher in the case of channel 3 due to the lower number of intermediaries involved. Specifically looking into channel 1, wherein farm gate sales take place, the

Table 7.10 Price spread of coffee (₹/kg) in domestic markets (2016–17)

Particulars	*Robusta Cherry (₹/kg)*[5]			*Arabica Parchment (₹/kg)*		
	Channel 1 *P-A-Cu-R*	*Channel 2* *P-H-Cu-R*	*Channel 3* *P-Cu-R*	*Channel 4* *P-A-D-Cu-R*	*Channel 5* *P-D-R*	*Channel 6* *P-Cu-R*
Gross price received by producer/Net price paid by agent or huller or curer or depots (A)	58.20 (19.40)	59.00 (19.66)	62 (20.66)	166.50 (37.00)	168 (37.33)	169 (37.55)
Costs incurred by producer (B)	1.40 (0.46)	1.40 (0.46)	1.80 (0.60)	1.50 (0.33)	1.80 (0.40)	1.90 (0.42)
Net price of producer (*C* = *A* − *B*)	56.80 (18.93)	57.60 (19.20)	60.20 (20.06)	165 (36.66)	166.20 (36.93)	167.10 (37.13)
Gross price received by agent /Net price paid by curer or depot (*D*)	63.00 (21)	-	-	168.50 (37.44)	-	-
Costs incurred by agent (*E*)	0.60 (0.20)		-	0.50 (0.11)	-	-
Net margin of agent [(*F* = *D* − (*A* + *E*)]	4.20 (1.40)	-	-	1.50 (0.33)	-	-
Gross price received by depot /Net price paid by domestic roasters or curer (G)				171.90 (38.20)	171.60 (38.13)	-
Costs incurred by depot (I)				2.90 (0.64)	2.90 (0.64)	-
Net margin of depot [(*J* = *G* – (*D* + *I*)]				0.50 (0.11)	0.70 (0.15)	-
Gross price received by huller /Net price paid by curer (*K*)	-	63 (21)	-	-	-	-
Costs incurred by huller (*L*)	-	2.16 (0.72)	-	-	-	-
Net margin of huller [(*M* = *K* − (*A* + *L*)]	-	1.84 (0.61)	-	-	-	-
Gross price received by curer/ Net price paid by domestic roaster cum retailer (*N*)	72.00 (24)	68.50 (22.83)	70 (23.33)	183.32 (40.73)	-	183.32 (40.73)

(Continued)

Table 7.10 (Continued)

Particulars	*Robusta Cherry (₹/kg)*[5]			*Arabica Parchment (₹/kg)*		
	Channel 1 *P-A-Cu-R*	*Channel 2* *P-H-Cu-R*	*Channel 3* *P-Cu-R*	*Channel 4* *P-A-D-Cu-R*	*Channel 5* *P-D-R*	*Channel 6* *P-Cu-R*
Costs incurred by curer (*O*)	3.73 (1.24)	3 (1.00)	3.73 (1.24)	2.70 (0.60)	-	2.70 (0.60)
Net margin of curer (*P*)	5.27 (1.75)	2.50 (0.83)	4.27 (1.42)	8.72 (1.93)	-	11.62 (2.58)
Gross price received by domestic roasters cum retailers (*Q*)	300 (100)	300 (100)	300 (100)	450 (100)	450 (100)	450 (100)
Costs incurred by domestic roaster cum retailer (*R*)	102.84 (34.28)	99.97 (33.23)	101.20 (33.73)	194.12 (43.13)	185.51 (41.22)	194.12 (43.13)
Net margin of domestic roaster cum retailer S = [(*Q* − (*N* + *R*); *S* = [*Q* – (*G* + *R*)	125.16 (41.72)	131.53 (43.84)	128.80 (42.93)	72.56 (16.12)	92.89 (20.64)	72.56 (16.12)
Selling price of roasters cum retailers (*T*)	300 (100)	300 (100)	300 (100)	450 (100)	450 (100)	450 (100)
Price Spread = *T* − *A*	241.80	241	238	283.50	282	281
Marketing costs (MC)	108.57	106.53	106.73	201.72	190.21	198.72
Marketing margin (MM)	191.43	193.47	193.27	248.28	259.79	251.28
Acharya's method of modified marketing Efficiency (MME) = [FP/(MC + MM)]	0.19	0.19	0.20	0.37	0.37	0.38

Source: Own compilation.
Note: Figures in parentheses indicate per cent to selling price of roasters cum retailers.

efficiency index is almost similar to the index of channel 2 and channel 3, wherein the coffee was sold to hullers (0.19) and curers (0.20).

Likewise, in the case of Arabica parchment, there does not arise significant differences in the marketing index when sold to agents at the farm gate or sold to curers or at coffee depots. The efficiency index when sold at the farm gate is almost similar to the one sold at depots or to curers or to the hullers. Therefore, the hypothesis that marketing efficiency is least in channel I (i.e., farm gate), while it improves in channel II and channel III (viz., hullers, curers, coffee depots and so on) cannot be accepted as the efficiency index does not improve substantially when sold to hullers, curers or depots.

7.5 Factors affecting net prices of coffee growers (Y = net price (₹/kg))

The net price (₹/kg) received by the growers was regressed on the area under coffee (ha), moisture content (%), out-turn (%), total assets (₹), and cost of harvesting (₹/ha). The functional form of the model is given below.

$$\text{Ln } Y \text{ (price/kg)} = \alpha + \beta_1 \ln \text{(area under coffee)} + \beta_2 \ln \text{(moisture content)} + \beta_3 \ln \text{(out-turn)} + \beta_4 \ln \text{(total assets)} + \beta_5 \ln \text{(harvesting costs)} + e$$

where Y = Price/kg

α = Constant or intercept
e = unexplained error term

Table 7.11 indicates factors affecting the net price received by the growers. The five factors, viz. area under coffee (ha), moisture content (%), out-turn (%), total assets (₹) and cost of harvesting were included in the model. In the case of Arabica, the value of R^2 was found to be 0.44. Here the moisture content was found to be negative and out-turn was positive and significant. This means that every 1 per cent increase in the moisture content reduced the net prices by 0.09 per cent, and every 1 per cent increase in the out-turn increased the prices of Arabica by 0.75 per cent.

In the case of Robusta, the area under coffee was negative and significant at the 5 per cent. In other words, for every 1 per cent increase in the area under coffee, prices reduced by 0.02 per cent. Every 1 per cent increase in the moisture content of Robusta decreased the prices of coffee by 0.17 per cent. The out-turn positively influenced the price; for every 1 per cent increase in the out-turn (%) of Robusta, the prices increased by 0.43 per cent. For every 1 per cent increase in the harvest cost, the prices of Robusta increased by 0.21 per cent, which was found to be significant at the 1 per cent level of significance. The total assets owned by the Robusta growers were found to be significant at 10 per cent and were positive. Overall, it can be observed that higher quality coffee (moisture and out-turn) is associated with a higher price.

Table 7.11 Factors affecting the net price (₹/kg)6received by the Robusta and Arabica growers (Y = ₹/kg)

Variables	*Arabica (N = 60)*			*Robusta (N = 60)*		
	Coefficient	*t stat*	*p value*	*Coefficient*	*t stat*	*p value*
Constant	1.98	2.53	0.01	2.50	3.36	0.001
Ln area under coffee (ha)	0.0007	0.18	0.85	−0.02	−2.16**	0.034
Ln Moisture content (%)	−0.09	3.18***	0.002	−0.17	−2.10**	0.039
Ln Out-turn[7] (%)	0.75	4.36***	5.7E-05	0.43	1.97**	0.050
Ln Harvest costs (₹/ha)	0.017	0.48	0.63	0.21	4.43***	4.54E-05
Ln Total assets (₹)	−0.003	-0.51	0.60	0.04	1.89*	0.06
.	R^2 = 0.44 *F* ratio = 8.55***			R^2 = 0.44 *F* ratio = 8.76***		

Source: Own compilation.
Note: ***, ** and * significant at 0.01, 0.05 and 0.10 level, respectively.

Currently, the traders, while purchasing coffee from the growers, check for moisture and out-turn percentage that determine the quality of beans. Regarding moisture, coffee with too high or too low moisture does not maintain a high cupping quality and is prone to mould contamination. Since these two are the major quality parameters checked by the traders, we analyzed the influence of moisture and out-turn on quality. An increase in the moisture content reduces the prices of coffee, as shown in Table 7.11.

Likewise, the out-turn or quality of seeds also determines the prices received by the growers. Here, the out-turn depends on the variety, elevation of the region and on the agronomic practices followed on the farm. As shown in Table 7.11, a higher out-turn represents a higher quality of beans, in turn reflected by higher net prices received by the growers. Here, out-turn and moisture content are the two parameters that necessarily determine the quality of coffee, wherein lower moisture content and higher out-turn represent better quality. Therefore, the hypothesis that there exists a strong and positive relation between net price realization and the quality of coffee in terms of out-turn and moisture content can be accepted.

7.6 Problems faced by producers in maintenance of quality

7.6.1 Genetics (species and genotypes)

The quality of coffee surfaces from the quality of plant varieties. Generally, it is accepted that the Arabica species of coffee is of superior quality compared to Robusta in terms of organoleptic qualities. The Robusta (caffeine content 1.5 to 2.5 per cent) with the stronger and harsher taste and twice the caffeine content of Arabica (1 to 2 per cent caffeine content) is found to be more suitable for inexpensive instant coffee. In terms of bean size and shape, the Arabica variety has a slightly larger shape than the smaller-sized Robusta. The Arabica coffee is also characterized by 60 per cent higher lipid content and chlorogenic acid (CGA). Even the sucrose content in Arabica is almost double (6 to 9 per cent) that of Robusta (3 to 7 per cent). For these reasons, Arabica is considered to be of superior quality, demanded by the top roasters with a higher price tag and used in the preparation of gourmet coffee due to its quality-related traits.

However, as discussed earlier, coffee growers are shifting from the Arabica to Robusta variety due to the susceptibility of Arabica to increase in temperature and precipitation, pests and disease attacks and higher costs of management. Such shifts could severely affect the quality of the coffee, as Robusta coffee is globally considered to be of lower quality grown at lower altitudes. With this, the country would lose competitiveness in the global markets in terms of quality.

7.6.2 Sustainability of coffee initiatives

Though certified coffee promises higher quality and sustainability standards, very few coffee producers have ventured into taking certification initiatives due to several complexities. Firstly, there exists low awareness among growers with regard to sustainable coffee initiatives, while the information sharing (usually undertaken by the NGOs and corporates) in this regard is also being limited. Second, the extra costs incurred in certification. As Ellis and Keane (2008) observed, the certification costs incurred by the small growers were US $500, while the large growers ended up paying much higher amount. However, the price premiums obtained for higher quality do not compensate the higher costs of certification, except for the Fairtrade certification, which guarantees a minimum price premium of US $1.25 per pound of coffee. However, Rainforest Alliance (RFA) and Common Codes for Coffee Community (4C) do not guarantee any such premiums for higher quality. Besides, there is no guaranteed market for certified coffee in India.

While in the case of organic certification, Chengappa et al. (2016) noted that more than 50 per cent of the surveyed organic coffee growers switched back to the conventional method of coffee production owing to the difficulties involved in obtaining organic inputs and the extra burden (record keeping), higher costs of organic certification with decreased productivity (30 to 50 per cent) during the transition period, and the price premiums being generally low viz., ₹200 for a bag of 50 kg. Meanwhile, Upendranath et al. 2012 observe that although the certification guidelines are transparent and easily understood by the growers, they do not comply with the local context and ground realities of Kodagu. On the other hand, different certification standards have created confusion in the minds of the growers. The growers opine that, rather than sustainability initiatives, the companies are interested in enhancing their brand image through certification schemes. Moreover, several certifying agencies promote different certification initiatives, which have created uncertainty among the growers.

7.6.3 Coffee Berry Borer

Among the major pests of coffee, the Coffee Berry Borer (*Hypothenemus hampei*) attacks berries, resulting in weight loss and a low out-turn percentage, which affects the price of the coffee. It is estimated that the weight loss is as high as 55 per cent in berry-attacked beans, while Magina (2005) reported crop losses as high as 96 per cent in heavily infested African countries. Studies conducted by the Central Coffee Research Station (Chetalli, Kodagu) in 2001 indicated the loss in clean coffee according to the degree of infestation. When the level of infestation was 10 per cent, the loss in clean coffee was 2.7 per cent. However, in cent per cent borer infestation, the loss in clean coffee was 37.20 per cent (Figure 7.8). The conversion ratio at the processing stage drastically reduced to <17.1 per cent, else under normal conditions, the

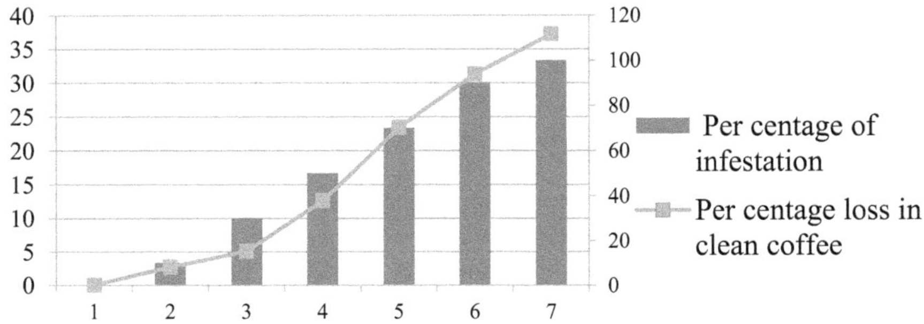

Figure 7.8 Loss in clean coffee at different levels of Berry Borer infestation (%)

Source: Bentley et al., 2002.

conversion rate is 5:1 (Baker et al., 2002). Therefore, it fetches lower prices for Berry Borer-infested beans.

7.6.4 Non-availability of resistant cultivars

Although the breeding effort led to the development of improved selections like Coorg, Kent, S 795 and Chandragiri, the exploitation of spontaneous hybrids and other diploid species resulted in the evolution of several cultivars. Identified DNA markers linked to disease-resistant genes, however, were found to be less successful in controlling leaf rust (*Hemileia vastatrix*), Coffee Berry disease (*Hypothenemus hampei*), Coffee White Stem Borer (*Xylotrechus quadripes*) and nematodes (*Meloidogyne* spp and *Pratylenchus* spp). There have been limited improvement efforts in this regard, particularly for the Robusta variety. The extension activities were also low in sourcing the improved varieties to the growers.

7.6.5 Rule 7B of Income-tax Act, 1962

The Rule 7B states, "Income derived from the sale of coffee grown and cured by the seller in India shall be computed as if it were income derived from business, and twenty-five per cent of such income shall be deemed to be income liable to tax". This tax law had disadvantaged a vast majority of the growers who sell their produce at the farm gate without any processing or value addition, which deteriorated quality of coffee and reduced the earnings of the growers, particularly marginal and small growers. Besides, the rule had further discouraged the growers from auctioning through the Indian Coffee Trader Association (ICTA). Instead of benefiting the growers for value addition, it has, in contrast, taxed the growers.

7.6.6 Harvesting and labour problems

This problem was mainly related to the type of harvest followed in coffee that determines the quality of beans. It could be observed from Table 7.21a that about 71.67 per cent of Robusta growers resorted to strip harvesting of coffee, while only about 28.33 per cent of Robusta growers adopted both methods of harvest. In contrast, about 83.33 per cent of Arabica growers adopted both methods of harvest. Strip harvesting involved the collection of both ripened and unripened berries at different levels of maturation, resulting in non-uniform drying, higher percentage of defects such as blacks, browns, bits and Berry Borer-affected beans and a higher percentage of fallen berries. These fallen berries, if not gleaned properly, could lead to Berry Borer infestation. Besides, the weight loss in roasted coffee beans was relatively higher in the case of strip-harvested coffee (12.13 per cent) compared to selectively harvested beans (14.42 per cent). On the other hand, selective picking involved only picking fully ripened cherries by hand, which ensured highest quality especially used for specialty coffee. However, Robusta growers do not adopt selective harvesting due to economic reasons. For instance, it cost about ₹3.78 per kg for selective and strip harvesting of coffee, while it cost merely ₹2.42 per kg for strip harvesting of Robusta. Firstly, selective picking of coffee was a laborious, time-consuming process, and several complexities were encountered for selective picking of coffee in hilly regions.

Secondly, gleaning was found to be less common among the surveyed growers. About 57 per cent of the Robusta and 43 per cent of Arabica growers undertook gleaning activities post-harvest. The major problem was the availability of workers (especially women) for the gleaning of coffee. Immediately after harvest, the female workers were engaged in drying, racking, weighing of coffee and cleaning and bagging of coffee, while the male workers were engaged in other intensive activities such as pruning, blossom irrigation, pre-blossom manuring and fungicide spraying. The growers thereby faced an acute shortage of labour for gleaning after harvest. Additionally, in the case of Robusta growers, about 87 per cent of them used harvest mats, wherein the chances of fruits falling onto the ground were meagre during the process of harvest and that the fallen berries were collected from the mats and not in contact with the soil. The growers also opined that the percentage of left-overs in their estates post-harvest was negligible and preferred not to proceed with the gleaning activities, instead diverting labour for other operations. The survey revealed that the growers were not very keen on undertaking the gleaning activities. Likewise, Achoth (2005) noted that gleaning was not very popular in the coffee-growing regions of Karnataka except Hassan.

7.6.7 Technical barriers

The use of collection or harvest mats as a phytosanitary measure against Coffee Berry Borer was popularized by the Coffee Board during the 1990s

(Achoth, 2005). During the harvest of coffee, harvest mats were used to prevent direct fruit contact with soil, thereby reducing the number of gleans on the ground, which were potential sources of pests, black beans, moulds and mycotoxins on the beans.

(i) As seen from Table 7.21a, the harvest mats (tarpaulins or polythene sheets) were used by 87 per cent of the Robusta growers. This practice was less commonly practised among the small farmers (50 per cent), whereas 88 per cent of semi-medium growers used this practice, while only 43 per cent of Arabica growers used harvesting mats. However, the growers' opinion was that the use of tarpaulins as ground mats was more expensive and less durable, which were the major reasons for non-adoption of mats during harvest.
(ii) In the case of processing, only about 6.67 per cent of the Robusta growers wet processed their coffee to produce parchment (Table 7.21a). The wet-processed Robusta beans appeared soft with neutral tones and a bluish-grey colour that represented higher-quality beans used for the preparation of espresso coffee. However, the dry-processed Robusta produced a harsher taste. The barriers to the preparation of parchment coffee were attributed to the complications and extra risks involved in wet processing Robusta due to the size and shape of Robusta beans compared to Arabica; Firstly, its thicker and adhesive mucilaginous cover (David et al., 2008). Secondly, the fermentation process in Robusta could not be completed, even if kept for a very lengthy fermentation period of 72 hours. Thirdly, wet-processed coffee required selective harvesting wherein only ripened berries were used. However, as discussed earlier, the procedure for wet processing was considered to be expensive and required higher power and water consumption.
(iii) About 78 per cent of Robusta growers mixed the ripened and unripened berries, which was found to be more prominent among the small growers (100 per cent), followed by semi-medium (88 per cent) and large growers (75 per cent) (Table 7.21a). Generally, the unripened beans were considered to produce poor cup quality and, when mixed with fully ripe berries, led to deterioration of the overall quality of beans. The mixing of ripened and unripened berries occurred more prominently when the coffee was strip harvested with non-uniform ripening of beans and due to non-sorting of beans.
(iv) Fourthly, about 85 per cent of Arabica and 90 per cent of Robusta growers heaped berries before drying, which varied between 1 and 5 days. However, such heaping of coffee led to mould formation. The major constraint for such practice was the non-availability of open space for drying owing to smaller drying yards. On the other hand, the growers believed that the practice of heaping was followed so as to rupture the fruit skin, which enabled faster drying of fruits (Achoth, 2005).

(v) It could be observed that about 6.67 per cent of Arabica and 10 per cent of Robusta growers used mud/cow-dung drying yards. The coffee, being hygroscopic in nature, absorbed the surrounding atmospheric moisture, with mould formation in the cherry husk and risks of OTA contamination found to be higher on the mud/cow-dung drying yard. Additionally, about 13.33 per cent of Arabica and 20 per cent of Robusta growers do not own separate storage facilities (Table 7.21a).

7.7 Problems faced by traders and exporters

7.7.1 Limited repute of Indian coffee in the global market (relative to the highest reputation regions)

Though Indian coffee enjoys a good reputation in the European market, it does not come under 'coffees of highest global repute' such as Jamaican Blue Mountain coffee (quality due to elevation between 914 and 1676 m.a.s.l. with unique growing conditions and volcanic soils), Hawaiian Kona coffee (quality due to rich volcanic slopes that were characterized by sunny mornings, cloudy afternoons and mild nights), Ethiopian Harrar coffee (elevation between 1371 and 1920 m.a.s.l. and dry processed), Mocha Java coffee (Blend of Arabian Mocha and Indonesian Java coffee) or Tanzania Peaberry coffee (grown under unique climatic conditions on Mt. Meru and Mt. Kilimanjaro).

Unlike the majority of coffee-growing regions across the globe, Indian regions cannot boast of volcanic soils, consistent rainfall patterns, temperature and unique blends of coffee from distinct origins. Though the Arabica variety from Bababudangiri is being cultivated at an elevation between 1000 and 1500 m.a.s.l., known to be of top quality similar to a 'Terroir' of coffee, its reputation in the global market remains very limited (David and Clementine, 2008). Moreover, Indian coffee does not enjoy a top 10 quality repute. For instance, coffee industry experts ranked Ethiopian coffee first in terms of quality, followed by Kenya, Colombia, Guatemala and Honduras, as shown in Figure 7.9.

7.7.2 Contractual problem

Generally, the coffee exporters entered into general agreements (contracts) with their clients for proposed transactions on the future exchanges, in which the quality, quantity, price, shipment period and conditions for sales were stipulated well in advance. However, inevitable complications (problems in sourcing adequate quantity for exports) do occur as against the prior contract descriptions. In order to fulfil their prior commitments, the exporters end up sourcing coffee from additional sources (curers or agents) and from different origins, without any geographical identification.[8] At such instances, the quality dilutions are likely to occur (geographical identification and differences in the quality of beans due to processing and handling). Thus, differences

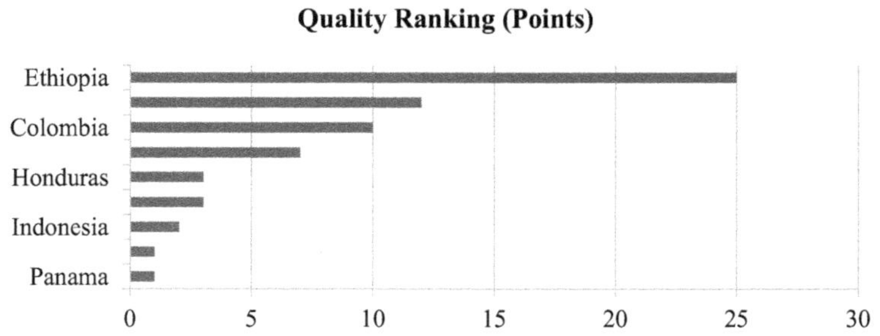

Figure 7.9 Rankings of coffee from different countries.

Source: Gentile, 2014.

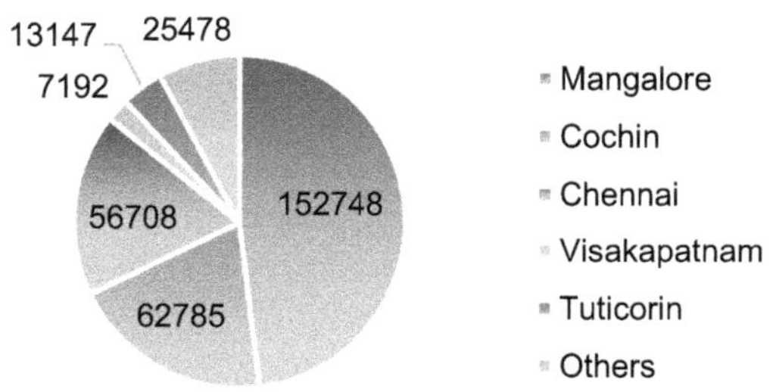

Figure 7.10 Quantity (MT) of shipment by various Indian ports (2015–16).

Source: Indian Coffee Board, 2017.

in cup tasting arise due to differences in physical and chemical complexities. Although the shipments are Subject to Approval of Sample (SAS), the situation could be complicated if the buyer's quality specifications are highly discrete, and they also end up buying coffee from other origins.

7.7.3 Uncertainty in weather conditions

About 48.03 per cent of Indian coffee was being exported from Mangalore port, followed by Cochin (19.74 per cent) and Chennai port (17.83 per cent) (Figure 7.10) to countries like Japan, Singapore, Cyprus, Kuwait, Italy, Switzerland, Israel, Germany, Belgium, Jordan and so on. These coastal ports were often affected by adverse weather conditions, viz. uncertainty in rainfall and tropical cyclones that in turn affected the quality of coffee during exports. The coffee, being hygroscopic in nature, easily absorbs the atmospheric humidity that further facilitates the growth of bacteria and fungi infestations, besides resulting in delayed shipments.

7.7.4 *Packaging material*

Another major problem faced by the exporters was obtaining the right packaging material at affordable prices that could play a significant role in maintaining export quality. A majority of Indian exporters usually packed the green coffee in woven natural materials like jute or sisal bags that allowed free air circulation. Though the jute bags were cost-effective and easily available, they did not provide necessary protection for the green coffee against moisture or other potential contaminants due to their porous open weave. Moreover, the hessian bags of jute or sisal were problematic during cargo transit when there were changes in the environment and relative humidity. Due to the hygroscopic nature of coffee, it absorbs the surrounding moisture, resulting in weight gains at the point of destination. This could particularly arise during the third and fourth quarters of the year when about 42 per cent of the coffee consignments were being exported (Table 7.22a). The exporters noted that usually the net weight when recorded during packing was 60 kg, but at the port of destination, the weight gain would likely be 69 kg. This could further encourage mould formation and undesirable flavour during exports. Moreover, hydrocarbon contamination was higher when packed with jute bags due to the 'batching oil' applied to smoothen the jute fibres prior to the spinning process. However, at times coffee could be shipped in an inner perforated plastic bag covered by the outer sisal bags, which could help in the protection of the quality of coffee to a certain extent.

The exporter's opinion was that the tin could prove to be an excellent packaging material for exports, serving as an excellent barrier for water, oxygen and light. However, there were three major issues with tin: firstly, the non-availability of an adequate number of tin containers in the country; secondly, the high expense of tin packing material; and, thirdly, the tin boxes held only a small quantity of coffee unlike the jute bags. While countries like Brazil exported their coffee in tin containers, the bulk package in plastic-lined sheets proved to be beneficial as it maintained the uniform temperature and moisture of the beans.

7.7.5 *Coffee beetle* (Araeocerus fasciculatus)

Another major problem faced by the coffee exporters was the presence of the coffee-bean weevil (*Araeocerus fasciculatus*), a storage pest that spread during the time of storage ashore. The beetle larvae dug inside the beans and pupated inside, while the adults emerged by boring a hole. Though the coffee-bean weevil was considered a minor pest of stored products, it could pose significant damage to the beans above 28°C; however, its existence in the export consignment could result in the rejection of the entire lot as the buyers believed that insect-damaged beans could cause tainted liquors. In countries like Iran, Egypt, Guatemala and Panama it was considered a harmful weevil.

7.7.6 OTA contamination during shipment

In coffee, Ochratoxin-A (OTA) is a naturally occurring toxin known as mycotoxin that is produced by moulds such as *Aspergillus, Penicillium and fusarium species.* OTA-contaminated beans have been reported to have a toxigenic nature and are known to damage kidney (nephrotoxic impact). OTA had always been a concern for exporters, primarily due to the hygroscopic nature of coffee, wherein the moisture content absorbed during shipping encourages mould contamination. Additionally, importing countries have imposed maximum permissible limits for OTA in the green, roasted and instant coffee that range between 3 and 20 ppb, as shown in Table 7.23a. However, excess levels of OTA would be subject to rejection due to the difficulty of eliminating OTA even at the roasted stage, despite green coffee being subjected to high temperature.

7.7.7 Difficulty in sourcing sustainable or certified coffee

In India, coffee certification is still at an infancy stage despite the well-established certification programmes across the globe (Upendranath and Subbaiah, 2012). The certified coffee represented Good Agricultural Practices (GAP) that contributed to the production of high-quality coffee beans, thereby brewing a better taste. However, the share of India towards global sustainable production was merely 2 per cent during 2012–13 (Figure 7.11), mainly sourced from large estates and corporations. Though there were initiatives such as group certification for small growers supported by exporters, the growers ended up selling coffee at the mainstream market itself owing to dissatisfaction with coffee prices. This had further discouraged the exporters from investing in training, certification or verification services for the growers.

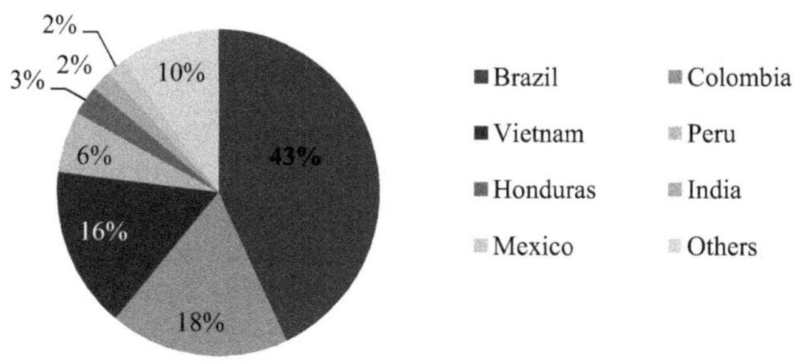

Figure 7.11 Country-wise share of sustainable coffee production in 2012–13 (%).
Source: Potts et al., 2014.

7.7.8 *Heavy dependence on agents and middlemen for procurement of coffee*

The exporters/curers were highly dependent on agents or middlemen who collected coffee from remote villages or farm gates and provided them with the required quantities. At the farm gate, the agents or middlemen looked for quantity rather than quality. The three major problems of coffee sourced from these agents included the moisture content, blacks, broken beans and physical impurities such as stones, leaves and twigs. Moreover, the agent middlemen were not professionally qualified to check for quality. The hullers were particularly blamed for the deterioration in quality. The coffees sourced from these agents were required to be re-dried, re-processed and classified into different quality grades. However, the exporters noted that even after re-processing the coffee, certain impurities still remained in the lot, which could be due to inadequate modern technologies required for processing.

7.7.9 *Lower value addition among exporters*

The major impediment for exports of Indian coffee was the lack of value addition at different stages of the market channel. As shown in Figure 7.12, the maximum value addition takes place at the roasters' level (degree of roasting, grinding, processing, specialized blending, brewing and cooling process) and at the retailer level (packaging, branding, product display advertisement), while merely 7 per cent is added at the exporters level.

Additionally, the proportion of value-added coffee (instant, ground and roasted coffee) exported from India was relatively low. For instance, the exports of value-added coffees such as roasted and ground coffee during 2016–17 merely formed 0.02 and 0.11 per cent of total coffee exports (Table 7.24a). However, the exports of value-added instant coffee remained relatively high, at around 28.91 per cent of total exports, due to high export demand from countries such as Russia, Turkey, Poland, Malaysia and so on.

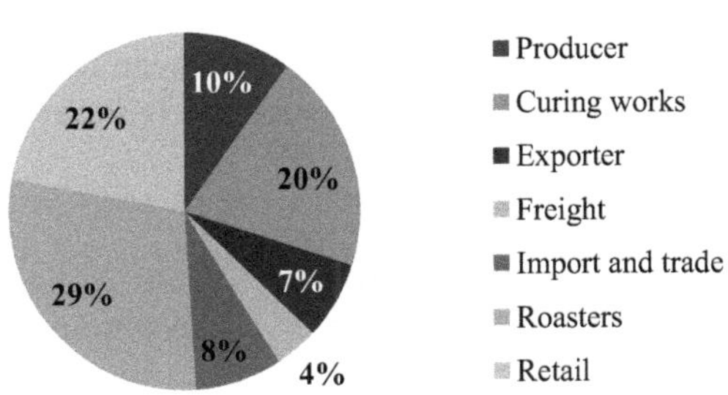

Figure 7.12 Percentage of value added on the price of dry processed coffee.

Source: Retrieved from David et al., 2008.

As mentioned above, the per cent of value-added coffee constituted a minor portion of the total Indian exports. However, there were several constraints for lower exports of value-added coffee (especially roasted and ground coffee). Firstly, the exporters were not able to produce the exact blend composition preferred by the consumers (especially Western consumers). This 'blend composition problem' of exporters provided the roasters in the importing countries a highly competitive advantage over the exporters. Moreover, the roasters could also source coffee from other origins in order to provide distinguished blends to the consumers.

Secondly, the roasted coffee, if not properly vacuum packed, absorbed the environmental elements that resulted in oxidation and loss of flavour. Thirdly, the importing countries applied tariffs on value added products, unlike the green coffee exported from India. Another major issue was the existence of GST for value added products, viz. 12 per cent for roasted coffee and 28 per cent for instant coffee that could also prove hindrance for value addition in coffee.

Even the exports of specialty coffee were relatively low, viz. about 16,237 MT in 2016. The specialty coffee (Monsooned Malabar AA, Monsooned Robusta AA, Robusta kapi Royale, Mysore nuggets EB and Monsooned Basanally) were well received in the global markets, specifically in the United States, European Union, Canada, Australia and New Zealand.

7.7.10 Pesticide residues in Indian coffee

The stricter food safety regulations acted as barriers to Indian coffee exporters. The regulations mainly took into account the pesticide residues in coffee, labelling of pesticides, packaging, storage and disposal of pesticides. The Codex Committee on Pesticide Residues (CCPR) established maximum residual limits (MRLs) for coffee as shown in Table 7.25a. Some amount of chemicals appears as residues even at the level of exports. The exporters were also perturbed over the higher dosage of Lindane and Chloropyriphos that growers used in excess amounts to control WSB in Arabica. Meanwhile, coffee-importing countries like Japan and the European Union had begun inspecting Indian coffee imports for toxic residues. The Coffee Board also issued the export license only after the coffee had been tested and analyzed by a certified laboratory in India for the presence of Lindane in the coffee beans.

7.7.11 Difficulties in shipment

The logistics-constrained shipment of coffee from Indian ports also acts as a constraint for Indian coffee exporters. The major shipment ports of Indian coffee were Mangalore (48.01), Cochin (19.73), Chennai (17.82), Visakhapatnam (2.26) and Tuticorin (4.13) ports. The Indian exporters also faced delays in shipments due to the lack of shipping technologies, inadequate port modernization facilities, cargo handling issues and seaborne traffic problems. During the peak export season, it could create port congestion and delays that also resulted in increased fees due to insufficient trucking capacity. The shipment delays resulted in the deterioration of the quality of beans, as they absorbed atmospheric moisture owing to the hygroscopic nature of beans.

7.8 Summary

The chapter attempted to understand the marketing structure and pattern of coffee in Kodagu and documented the major problems faced by producers, traders and exporters in maintaining quality. The chapter further discussed the fluctuations in farm gate prices, the ratio of farm gate prices to ICO indicator prices and mode of sales. The price spread of coffee, marketing efficiency and factors affecting the net prices of coffee are detailed, and lastly the major problems faced by producers, traders and exporters for quality maintenance are discussed. The maximum fluctuations in farm gate prices were observed in the first quarter and fourth quarter for Robusta and Arabica, respectively, which also coincide with the procurement season of coffee.

The augmented Dickey–Fuller (ADF) test shows that for all the five series, viz. Arabica Cherry, Arabica International, Arabica Parchment, Robusta Cherry and Robusta International, the level of price data was non-stationary, wherein the ADF does not reject the unit test hypothesis. However, stationarity was reached after the first difference, implying that the data integrated of order one, $I(1)$. In this case, the hypothesis was rejected at 1 per cent, indicating that all five series were integrated of order 1. Johansen's Trace and Max Eigen value indicated the non-existence of co-integration between Arabica Cherry and International markets, Robusta Cherry and Robusta International, Arabica Parchment and Arabica International, Arabica International and Robusta International and between Arabica Cherry, Arabica Parchment and Arabica International prices.

The Granger causality test results revealed that there existed a bi-directional flow of price information from Arabica International to Arabica Cherry and also from Arabica Cherry to Arabica International, from Arabica Parchment to Arabica International as well as from Arabica International to Arabica Parchment as well. On the other hand, uni-directional causality was revealed in the prices of Arabica International and Robusta International. The same was true for Robusta International and Robusta Cherry.

In the study region, it was revealed that the price per kg of Robusta sold to roaster-cum-retailer was ₹300 per kg, while the producers received a net price of only ₹56.80, ₹57.60 and ₹60.20 per kg in channel 1, channel 2 and channel 3, respectively. The expenses incurred by producers varied between ₹1.40 and ₹1.80 per kg of Robusta. In the case of Arabica, the selling price of roaster-cum-retailers was found to be ₹450 per kg. The growers received only ₹165, ₹166.20 and ₹167.10 in channel 4, channel 5 and channel 6, respectively. The overall marketing efficiency was higher in the case of Arabica Parchment (0.38) against Robusta Cherry (0.19).

In terms of major factors impacting the net prices received by the growers, the moisture content and out-turn percentage had a major impact on the prices received by the growers in both varieties. Additionally, the cost of processing significantly influenced the prices of Arabica, and area under Arabica positively influenced the net prices received by the growers.

The problems faced by the growers in the maintenance of quality were the species and genotypes of coffee, non-availability of resistant cultivars, problems of certification, rule 7B and technical barriers to quality maintenance. On the other hand, traders faced problems associated with the limited repute of Indian coffee (as compared to Jamaican Blue Mountain and Kona coffee), contractual problems, uncertainty in weather conditions, packaging material, pests and diseases, OTA contamination, heavier dependence on agents or middlemen, lack of value addition and difficulties in the shipment of coffee.

Notes

1 Rule 7B of Income tax: "Income derived from the sale of coffee, grown and manufactured by the seller in India, with or without mixing of chicory or other flavoring ingredient shall be computed, as if it were income derived from business and 40 per cent of such income is liable to tax".
2 Note: The optimum lag length was determined using the Akaike Information Criterion (AIC).
3 Likewise, model presented for AP-AC, RI-AC, RC-AC, AP-AI, RI-AI, RC-AI, RI-AP, RC-AP, and RC-RI.
4 Likewise, model presented for AP-AC, RI-AC, RC-AC, AP-AI, RI-AI, RC-AI, RI-AP, RC-AP and RC-RI.
5 P-A-Cu-R – Producers-Agents-Curers-Roasters-cum-retailers; P-H-Cu-R – Producers-Hullers-Curers-Roasters-cum-retailers; P-Cu-R – Producers-Curers-Roasters-cum retailers; P-A-D-Cu-R – Producers-Agents-Depots-Curers-Roas ters-cum-retailers; P-D-R – Producers-Depots – Roasters-cum-retailers ; P-Cu-R – Producers-Curers-Roasters-cum-retailers.
6 Dependent variable is net prices (₹/kg) received by the grower.
7 Out turn means 1 kg of beans from 5 to 6 kg of berries or cherries.
8 At such instances, traders blend different qualities of coffee, including coffee imported from Vietnam. The Vietnamese coffee was usually subject to quality suspicion.

References

Achoth, L., 2005, *Report on surveys on coffee holdings and coffee market chain in India in relation to mould contamination in coffee*. Submitted to Coffee Board of India, Bangalore and Food and Agricultural Organization of United Nations, Rome.

Baker, P.S., Jackson, J.A.F. and Murphy, S.T., 2002. *Natural enemies, natural allies*. Project completion report of the integrated management of Coffee Berry Borer

project, CFC/ICO/02 (1998-2002). The commodities press. CABI commodities, Egham UK and Cenicafe, Chinchina, Colombia.
Bentley, J.W. and Baker, P.S., 2002, *Manual for collaborative Research with smallholder coffee farmers*. CABI Commodities, Egham, 130 pp.
Chengappa, P.G., Rich, K.M., Muniyappa, A., Yadava, C.G., Ganashruthi, M.K., Pradeepa Babu, B.N., Shubha, Y.C. and Rich, M., 2014a. *Sustainability coffee certification in India*. Perceptions and Practices. NUPI Working Paper 830, Norwegian Institute of International Affair.
Chengappa, P.G., Rich, K.M., Muniyappa, A., Yadava C.G., Babu, P. and Rich, M., 2014b. *Promoting conservation in India by greening coffee. A value chain approach*. NUPI Working Paper 831, Norwegian Institute of International Affairs.
Chengappa, P.G., Devika, C.M., Muniyappa, A. and Pradeepa Babu, B.N., 2016, *The future for sustainable coffee production in India: Certified and organic coffee*. NRPPD Discussion Paper No. 54, Centre For Development Studies, Thiruvanathapuram.
David, M. and Clementine, V., 2008, *Coffee value chain and geographical indications in India: Origin, reputation and marketing of Indian coffees*. CIRAD, Agriculture Research for Development.
Ellis, K. and Keane, J., 2008. *A review of ethical standards and labels: Is there a gap in the market for a new 'Good for Development' label?* Overseas Development Institute Working Paper 297, Overseas Development Institute, London.
Gentile, D., 2014, The definitive top 10 coffee-growing countries in the world, ranked by experts. Available at https://www.thrillist.com/drink/nation/the-world-s-best-coffee-growing-countries-ethiopia-kenya-colombia-and-more
Granger, C.W.J., 1969, Investigating casual relations by Econometric models and cross-spectral methods. *Econometrica* 37(3): 424–438.
Indian Coffee Board. 2017, Database on Coffee, March 2017. Available at www.indiacoffee.org
Kohls, R.L. and Uhl, J.N., 1980., *Marketing of agricultural products*, Macmillan Publishing Co.Inc., New York, p. 589.
Magina, F.L., 2005, *A review of coffee pest management*. Tanzania coffee research Institute, Lyamungu, Tanzania. pp. 99.
Potts, J., Lynch M., Ann Wilkings, Huppe, G., Cunningham M. and Voora, V., 2014, *The state of sustainability initiatives review: Standards and green economy*. International Institute for Sustainable Development and London and the International Institute for Environment and Development.
Upendranath, C. and Subbaiah, C.A., 2012, *Small growers and coffee marketing-issues and perspective from the field*. NRPPD Discussion Paper 15, Centre For Development Studies, Thiruvananthapuram.

Chapter 8

The role of institutions in production, marketing and exports of coffee

8.1 Introduction

The term "institution" is defined distinctly in the literature. In general, institutions could be defined as the norms, rules, regulations, property rights and policies that structure human behaviour (Tulu, 2010). Broadly, institutions in this context of research would mean formal and informal regulations that govern the coffee industry. Box 8.1 briefly presents the studies covering the definition of institutions and their various aspects.

Box 8.1: Definition of institution in the literature

Major literature reviews	
Veblen (1934)	Institution is a particular type of social structure with the potential to change the agents involved in the process, purpose and references. Institutions work as the rules necessitated that are rooted in the shared habits of thoughts and behaviour through which society evolves.
Coase (1937)	Institutions are the legal, political and social systems that determine economic performance. These institutions ought to coordinate transactions, reduce transaction costs (costs of information, coordination, contracts, monitoring and negotiation) and enable agents to discover relevant prices.
Commons (1951)	Institutions are the collective actions in control, liberalization and expansion of individual actions. Here, the collective action includes both unorganized and organized structures, viz. family, corporation, trade unions, associations, reserve system and the state.
Williamson (1985)	Institutions are the modes of managing transaction costs, which include market, quasi-markets and modes of contracting.

DOI: 10.4324/9781003541721-8

Major literature reviews	
North (1989)	Institutions are the rules of the game and humanly devised constraints that shape human interactions.
Crawford and Ostrom (1995)	Institutions are the norms, rules and strategies governing human actions.
Hodgson (2006)	Institutions are the social rules that structure social interactions.
Ostrom (2010)	Institutions are the codes of behaviour that potentially reduce uncertainty, mediate self-interest and facilitate collective action.

Source: Own compilation of literature.

The institutions could be broadly classified as formal and informal institutions. While Scott (2015) noted the various governance structures, non-government organizations (NGOs), legal and legislative institutions, social safety nets, community organizations and associations in the frameworks of institutions. The national regulatory framework, plans, policies, organizations, extension services and access to credits were included under the formal institutions, while the informal institutions included the ways of thinking, acting, adaptability, forms of collaboration and degree of inclusiveness and exclusiveness. With this backdrop, the chapter reassesses the institutional arrangements of coffee and attempts to understand the formal and informal institutions involved in the coffee industry.

8.2 Land tenure system as an institution in Kodagu

The land tenure system as an institution is governed by a set of rules and regulations formulated by the society in order to regulate human behaviour in a particular direction. Nevertheless, this also determined the utilization of resources with respect to land (Tenaw et al., 2009), allocation of property rights within society, viz. rights to use, control, transfer land and other resources.

Taking into account the land tenure system as an institution, this section deals with two major private land tenures in Kodagu, viz. redeemed and unredeemed land tenures and attempts to understand their role in the production aspects of coffee. Under the Coorg Land and Revenue Regulation 1899, the redeemed owner holds tree rights or the right to harvest trees in his estates, provided the owner had made payment for timber values of trees. Whereas in the case of unredeemed land, the government holds tree rights, though the land is legally owned by the coffee grower. In this case, the coffee growers needed to obtain the permission of the Karnataka Forest Department for tree felling besides the unredeemed land owners were required to pay the prescribed seigniorage value to the Government for tree felling (Garcia et al., 2010), thereby restricting the tree rights in their own

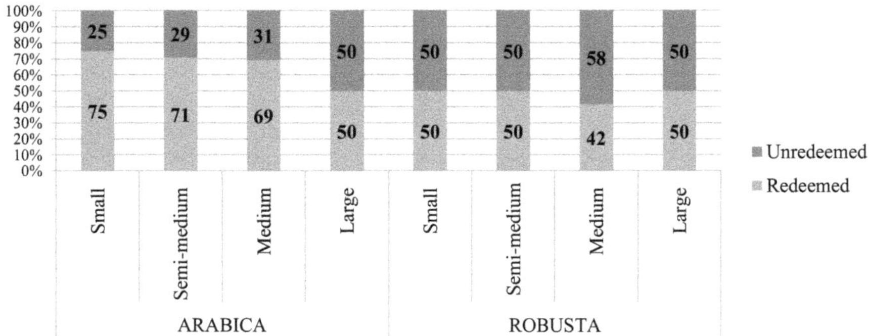

Figure 8.1 Redeemed and unredeemed land among the surveyed estates (%).

Source: Field survey.

estates (Uthappa, 2004). Here the institutions not only controlled individual behaviour (individual coffee growers were restrained from tree cutting) but also enabled action in a particular direction (conservation of native shade trees).

From our field survey, it was found that at an aggregate level, about 68.33 per cent of Arabica and 48.33 per cent of Robusta estates were redeemed, while about 31.67 per cent of Arabica and 51.67 per cent of Robusta estates were unredeemed (Figure 8.1). Interestingly, the redeemed estates were higher in the case of Arabica, about 75 per cent in small, semi-medium (71 per cent), medium (69 per cent) and large (50 per cent), while the Robusta estates comprised almost equal per cent of redeemed and unredeemed lands.

8.2.1 Land tenure system and productivity of coffee in Kodagu

The land tenure system impacts agriculture and resource management, as the secured property rights ensure sufficient impetus to the growers to enhance their productivity efficiencies while also ensuring economic sustainability (Tenaw et al., 2009). In this section, we attempt to link land tenure (redeemed and unredeemed land) with the productivity of the Arabica and Robusta varieties of coffee.

The productivity of the Arabica variety of coffee under different land tenure systems is shown in Figure 8.2. The productivity was higher in unredeemed land (813 kg/ha) than redeemed land (798 kg/ha). In the unredeemed land tenure, the productivity was highest among the large growers (825 kg/ha), followed by medium (812 kg/ha), semi-medium (811 kg/ha) and small (800 kg/ha). While in the case of redeemed land, productivity was found to

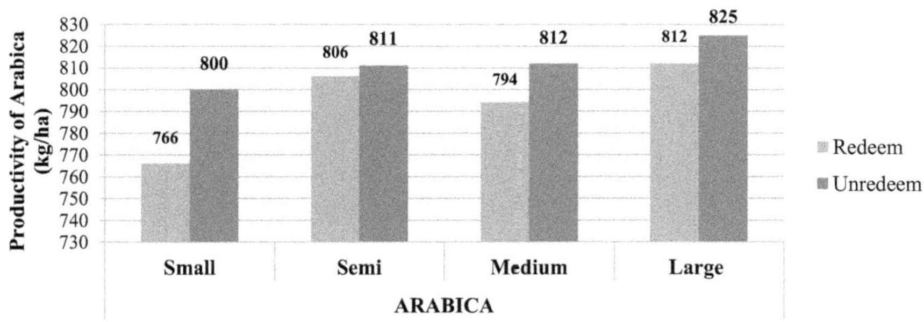

Figure 8.2 Productivity of Arabica (kg/ha) under different land tenure.

Source: Field survey.

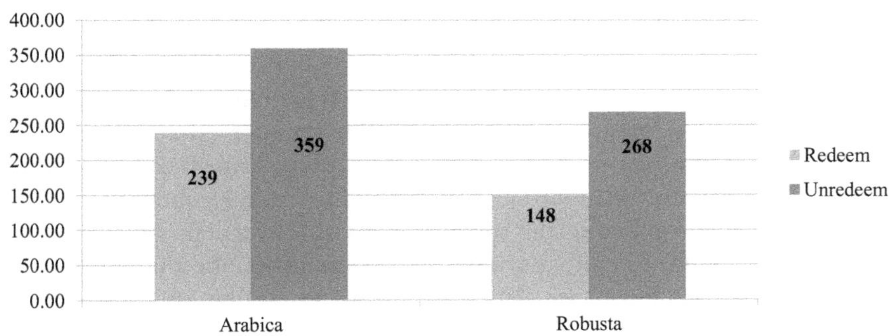

Figure 8.3 Number of shade trees per ha under different land tenure systems.

Source: Field survey.

be highest among the large growers (812 kg/ha), semi-medium (806 kg/ha), medium (794 kg/ha) and small growers (766 kg/ha).

As shown in Figure 8.3, the shade tree density was found to be higher in unredeemed Arabica estates (359 shade trees/ha) relative to the redeemed Arabica estates (239 shade trees/ha), as the former land tenure system restricted the growers from harvesting the native tree species. Likewise, Sathish et al., 2012) noted a higher tree density and volume under the unredeemed land tenure system. This higher shade tree density in unredeemed land provided optimum shade for the Arabica variety. The shade trees in Arabica estates also control the incidence of White Stem Borer (WSB) in Arabica. Thus, it could be concluded that the land tenure system (unredeemed land) had an impact on the productivity of the Arabica, as the control of tree rights by the forest department had compelled the growers to conserve indigenous

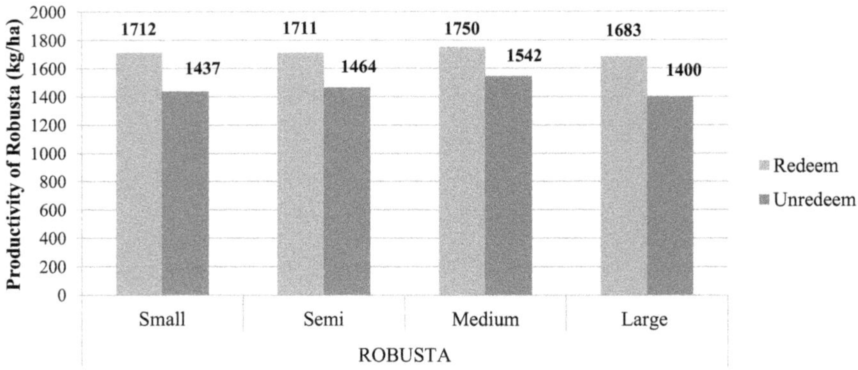

Figure 8.4 Productivity (kg/ha) of Robusta under different land tenure systems.

Source: Field survey.

tree species in their estates, thereby maintaining a higher shade cover in their estates that also had a positive impact on the productivity of Arabica.

Figure 8.4 shows the productivity of Robusta under different land tenure systems. In contrast to Arabica, the productivity of Robusta was found to be higher in redeemed estates. In this case, the tree rights conferred on the growers have resulted in the removal of shade cover. Specifically, in the regions where Robusta cultivation has been expanding, the shade trees and the indigenous trees have been removed. The lower number of shade trees in turn provides more sunlight into the estates. Meanwhile, the Robusta trees generally survive well at a higher temperature and humidity. Besides, the sunlight requirement is generally high in Robusta, which has necessitated photosynthetic activities in Robusta resulting in higher productivity. Unlike Robusta, Arabica is a shade-loving variety, traditionally grown under shade and thrives well under shade. The thick shade canopy is more pronounced in the unredeemed land, thereby providing a congenial environment for Arabica, reflected by higher productivity in unredeemed Arabica estates.

The overall productivity of redeemed Robusta estates was found to be 1715 kg/ha, much higher than the unredeemed land (1472 kg/ha). The Robusta productivity in redeemed estates was highest in the medium (1750 kg/ha), followed by small (1712 kg/ha), semi-medium (1711 kg/ha) and large growers (1683 kg/ha), whereas the productivity of Robusta in the unredeemed land was highest in the medium (1542 kg/ha), followed by semi-medium (1464 kg/ha), small (1437 kg/ha) and large (1400 kg/ha) (Figure 8.4).

8.2.2 *Land tenure systems in Kodagu and coffee cultivation practices*

8.2.2.1 Shade grown to sun grown (silver oak grown)

It could be observed from Figure 8.5 that a higher percentage of silver oaks were present in the redeemed land, viz. about 33 per cent and 62 per cent in redeemed Arabica and Robusta estates, respectively. In other words, the 'tree rights' conferred to redeemed estates had prompted the growers to plant exotic silver oaks (*Grevillea robusta*)[1] in place of indigenous native tree species. Furthermore, the loss of shade canopy was observed more in redeemed land tenure in both Arabica and Robusta, which in fact resulted in a major shift from traditional polycultures of native shade canopies to silver oak monoculture in coffee plantations (Damodaran, 2002).

Though the fast-growing exotic silver oak (*Grevillea robusta*) was preferred over the other native species due to its high timber value, it had adverse impact on coffee production. For instance Abraham et al. (2013) noted that the coffee bud emergence in the branches was adversely affected by the falling leaves of silver oaks that remained un-decomposed for several months. However, in the case of Arabica, the silver oak trees did not ensure optimum filtered shade, thereby inducing pests and diseases like White Stem Borer (WSB), Coffee Leaf Rust (CLR) and Koleroga. This reduced tree canopy in coffee estates had also resulted in increased temperature and reduced relative humidity, thereby making the environment uncongenial for cultivation of Arabica as this variety required higher shade cover and a lower temperature.

8.2.2.2 Shift from Arabica to Robusta variety of coffee under different land tenure systems

The shift from Arabica to Robusta variety of coffee under different land tenure systems is provided in Figure 8.6. In all cases, except among small

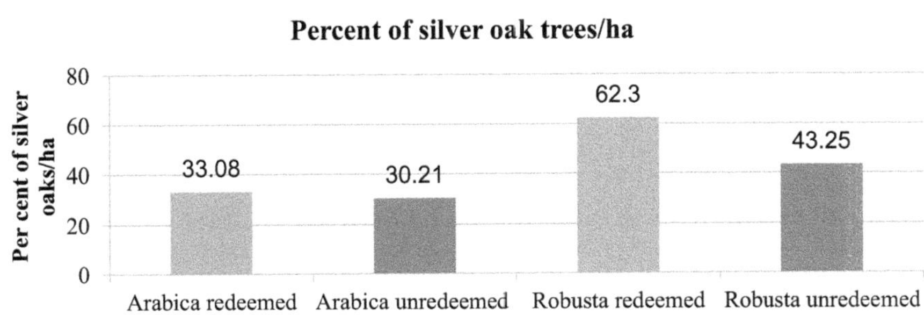

Figure 8.5 Percentage of silver oak trees under different land tenure systems. Per cent of silver oaks were calculated using (number of silver oak trees per ha/Total number of shade trees per ha) * 100.

Source: Field Survey.

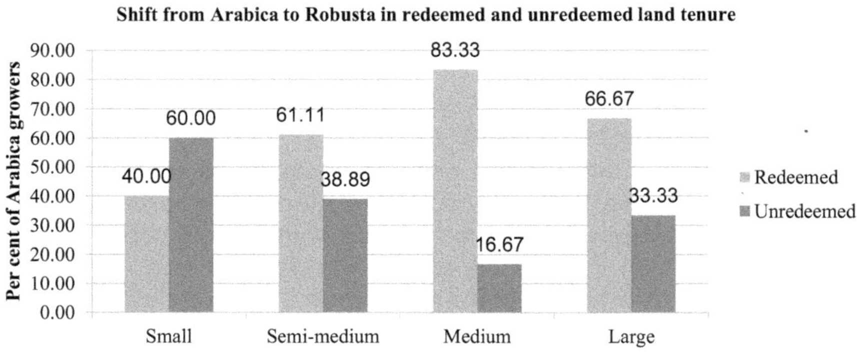

Figure 8.6 Shift from Arabica to Robusta variety under different land tenure.

Source: Field survey.

growers, the shift as expected was more pronounced in the redeemed land. About 83.33 per cent of the medium growers, followed by large (66.67 per cent), semi-medium (61.11 per cent) and 40 per cent of the small growers in redeemed type of land tenure were in the process of shifting from Arabica to Robusta coffee. On the other hand, the shift towards Robusta were relatively low under unredeemed land tenure except in the case of small growers, where about 60 per cent of growers were shifting towards Robusta, followed by semi-medium (38.89 per cent), large (33.33 per cent) and medium unredeemed Arabica growers (16.67 per cent).

Earlier, Arabica coffee was cultivated under the canopy of thick shade cover that comprised native tree species; however, these were now getting converted to Robusta, which required sparse shade (Garcia et al., 2010). Nevertheless, we observed that redeemed land comprised fewer shade trees (Figure 8.3) with a higher per cent of silver oaks (Figure 8.5). This in turn indicated that redeemed land comprised with its lower shade canopies and sparse shade cover was congenial for Robusta cultivation, which required lesser shade (40 per cent) as against Arabica, which required more shade (70 per cent) as evident in unredeemed estates.

In contrast to the above, the redeemed owners with available ownership of tree rights were allowed for the felling of indigenous trees which were being replaced with silver oaks that had maintained a lower shade in their estates (about 40–60 per cent), as required by Robusta. Moreover, the lower maintenance costs and ease of production of Robusta supported by the land tenure system (specifically in the redeemed land) had prompted the Arabica growers to shift towards Robusta cultivation which were more pronounced in the redeemed land (Figure 8.6).

Overall, the land tenure system in Kodagu as the norms and rules set by the government or as an institution had a modest impact on the coffee agro forestry system, productivity and coffee production practices. However, the current land tenure system was very traditional by itself, which was formulated 500 years ago by the then ruled Haleri dynasty of Kodagu, latter consented by the British government that persists even today in conflict with the interests of the land holders in Kodagu. Thus, in order to reconcile the 'traditional land tenure system' as an institution with the modern system, the government needed to execute a formal revenue survey which could also provide tenure security to the local coffee growers that could halt illegal encroachment of land for commercial coffee cultivation. This could also cease the proposals of the forest department and certain environmentalists to convert private coffee plantations into state-controlled forests by re-assigning them as 'deemed forests'. Further, there was also a need to re-look the tree standing rights in the private coffee plantations in interests of coffee growers in Kodagu.

8.3 The 'New Coffee Bill-2016' – Is it old wine in a new bottle? Presumptions and challenges

The encroachment of forest land for commercial coffee cultivation had been reported in the coffee-growing regions of Karnataka. This could be partly responsible for the non-accuracy of data on coffee cultivation, as some of the areas under coffee were not officially recorded. The official data provided by the Coffee Board were mere estimates that relied on the information provided by the extension workers and data obtained during the disbursal of subsidies.

Nevertheless, the Coffee Board of India along with the Indian Space Research Organization (ISRO) conducted a remote sensing mapping in the six coffee-growing regions of Karnataka from 2013 to 2015. The results revealed that the area under coffee was 14 per cent more than the officially recorded data (Businessline, 2017), which indicated a wide variation in the remote sensing data and officially recorded revenue data. The paramount reason for the non-accuracy of data could be the non-mandation of registration of the area under coffee.

In order to get the correct estimate of the area under coffee and production from individual estates, the Coffee Board seeks to consider compulsory registration of coffee growers through the introduction of the new Coffee Bill,[2] 2016. The section 15(1) in the Bill necessitated compulsory registration of the growers with the Coffee Board.[3] The Bill claimed that the main reason for registration was to obtain more accurate and reliable inventories of coffee-growing regions for the effective formulation and development of the coffee programmes.

Besides, the long-established Coffee Act[4] of 1942 still exists in the current scenario, while many of its provisions had become expendable specifically after the liberalization period. Therefore, the Union Government considers withdrawing the Coffee Act of 1942 and replacing it with the Coffee Bill of 2016. The new Act also empowered the Coffee Board with powers of registration (growers, curers, roasters and manufactures of instant coffee), formulation of strategies for the promotion of the coffee industry, market development including quality standards, issuing directives against major pests and diseases and to undertake research and development activities. Interestingly, the 'New' Coffee Bill 2016 appeared to be a mere replica of the Coffee Act 1942, except for one major difference, viz. the pooled marketing system followed earlier while the new bill followed the marketing of coffee through the private traders. The constitution of Coffee Board, functions and powers of the Board were the other minor changes to the new bill. Nevertheless, there were many similarities between the two acts which included preliminary, declaration, definition, consultation, registration of estates and cognizance of offences.

Box 8.2: Similarities between Coffee Act 1942 and Coffee Bill 2016[5]

S.No.	*Coffee Act 1942*	*Draft New Coffee Act, 2016*
1	**Preliminary:** An Act to provide for the development under the control of the Union, of the coffee industry: whereas it is expedient to provide for the development under the control of the Union, of the Coffee industry:	**Preliminary:** An Act to provide for the development **and regulation** under the purview of the Union, of the coffee industry: whereas it is expedient to provide for the development and regulation under the purview of the Union, of the coffee industry:
2	**Declaration as to expediency of Union Control:** It is hereby declared that it is expedient in the public interest that the Union should take under its control the coffee expediency industry.	**Declaration as to expediency of Union Purview:** It is hereby declared that it is expedient in the public interest that the Union should develop and regulate the coffee industry.

S.No.	*Coffee Act 1942*	*Draft New Coffee Act, 2016*
3	**Definition:** "Coffee" means the commodity derived from the fruit of rubiaceous plant known by that name, and includes raw coffee, cured coffee, uncured coffee, roasted coffee and prepared coffee.	"Coffee "means the **fruits (Cherries) and seeds (beans) of plants of the genus Coffea,** as well as products from these fruits and seeds in different stages of processing and use.
	"Curing" means the application to raw coffee of mechanical process other than pulping for the purpose of preparing it for marketing.	"Curing" means the application to raw coffee of mechanical processes other than pulping for the purpose of preparing it for marketing.
	"Dealer" means a person carrying on the business of selling coffee, whether whole-sale or by retail;	"Dealer" means a **person, firm or company** carrying on the business of coffee, whether whole-sale or retail;
	"Estate" means an area administered as one unit which contains land planted with coffee plants;	"Estate" means a **farm or contiguous area administered as one unit** of land used primarily for growing coffee.
	"Owner" in relation to any land planted with coffee plants, includes – (1) any agent of the owner, and (2) a mortgagee, lessee or other person in actual possession of the land.	"Owner" in relation to any land planted with coffee plants, includes – (1) any agent of the owner, and (2) a mortgagee, lessee or other person in actual possession of the land;
	Definition of 'Quality' is not defined in the 1942 coffee act.	**"Quality" means quality standards as applied to coffee beans in India or in the exporting countries or as per internationally recognized standards.**
	"Registered estate" means an estate in respect of which an owner is registered under sub-section (1) of section 14, and includes also any estate in respect of which an owner if required to be registered under the provisions of that Sub-Section."	"Registered Estate" means an estate in respect of which an owner is registered under sub-section (1) of section 15, and includes also any estate in respect of which an owner if required to be registered under the provisions of that Sub-Section.
	"Registered owner" means an owner of a registered estate who has been or is required to be registered under sub-section (1) of section 14.	"Registered owner" means an owner of a registered estate who has been or is required to be registered under sub-section (1) of section 15.

S.No.	*Coffee Act 1942*	*Draft New Coffee Act, 2016*
4	**Constitution of the Board:** **Chairman:** 1. Chairman to be appointed by the Central Government by Notification in the Gazette. **Members of Parliament:** Three members of parliament of whom two shall be elected by the House of the People and one by the Council of states 3.**Other members:** not exceeding 29 numbers as the Central Government may think expedient by notification in the Gazette from among: (ii) Coffee-growing industry; (iii) Coffee trade interests; (iv) Curing establishment; (v) Interests of labour; (vi) Interests of consumers; and (vii) Such other interests as in the opinion of the Central Government ought to be represented on the Board.	**Chairman:** 1. Chairman to be appointed by the Central Government by Notification in the Gazette **Members of Parliament:** Three members of parliament of whom two shall be elected by the House of the People and one by the Council of states 3. **Other members: not exceeding 20 numbers** as the Central Government may think expedient to be appointed by the Government by notification in the Official Gazette from among persons who are in its opinion capable of representing: - (i) **Governments of the Principal coffee growing States.** (ii) Coffee-growing industry; (iii) Coffee trade interests including marketing and development of coffee; (iv) Curing establishment; (v) Coffee Research; and (vi) Interests of labour; consumers and such other interest including environment, as in the opinion of the Central Government, ought to be represented on the Board.
5	**Consultation with the Board:** Before taking any action touching the affairs of the Board with the Board under this Act, the Central Government shall ordinarily consult the Board: Provided that no action taken by the Central Government shall be invalid or called in question merely on the ground that the action was taken without such consultation.	**Consultation (By Central Government) with the Board:** Before taking any action concerning the Coffee sector and functions of the Board under this Act, the Central Government shall ordinarily consult the Board: Provided that no action taken by the Central Government shall be invalid or called in question merely on the ground that the action was taken without such consultation.

S.No.	*Coffee Act 1942*	*Draft New Coffee Act, 2016*
6	**Election of Vice Chairman.** The Board shall elect from amongst its members a Vice Chairman who shall exercise such of the powers and perform such of the duties of the Chairman as may be prescribed or as may be delegated to him by the Chairman.	**Sec.11: Vice Chairman.** The Board shall elect from amongst its members a Vice Chairman who shall exercise such of the powers and perform such of the duties may be prescribed or as may be delegated by the Chairman.
7	**Sec.14. Registration of Coffee Estates.** Every owner of land planted with coffee plants, whether of owners of such land is comprised in one estate or in more than one estate and coffee whether it is situated wholly or only partly in India, shall, apply to the registering officer appointed in this behalf by the State Government	**Sec.15. Registration of Owners of Coffee Estates.** Every owner of land primarily planted with coffee plants, whether such land is comprised in one estate or in more than one estate apply to the Coffee Board to be registered as an owner in respect of each estate owned by him.
8	Functions of the Board do not exist under the Coffee Act 1942.	**Sec.7. Functions of the Board.** Formulating strategies, promotional activities, plant protection measures, quality standards, research and development, dissemination of information, advising central government, **Sec.18.**The Board shall have the power to inspect or get inspected the coffee being exported or imported to satisfy itself that such coffee is in accordance with the quality parameters prescribed by the Board
	Sec.40. Cognizance of offences: No court other than the Court of Judicial Magistrate of the first class shall take cognizance of any offence punishable under this Act.	**Sec 24. Cognizance of an offence** No Court shall take cognizance of an offence punishable under this Act except on a complaint made by an officer authorized in this behalf by the Chairman of the Board.

Source: Own compilation from literature.

However, the coffee growers and several organizations in Kodagu have strongly opposed the new bill. The growers assert that the Coffee Act 1942 was launched by the erstwhile British government during the Second World War to keep the coffee sector under its control, providing less liberty to the producers. With the enactment of the new bill, the growers fear that the system would bring back the regulated system into the coffee sector, controlled by the institutions and to a greater extent by the Coffee Board. However, the Coffee Act 1942 and Coffee Bill of 2016 were similar in many ways, except for a few minor changes.

Secondly, the registration aspect (Section 15) of the new Bill had created discontent among the growers, mandating compulsory registration of the growers. They presume that the Coffee Registration Certificate (CRC) would act as a tool for the exploitation of the growers, especially the marginal and small growers, by imposing registration fees, who thereby presume that they are liable for tax payment in the future, though the agricultural income is exempt from income tax.

Another major discontent among the growers was Section 26 of the New Bill, 2016, which empowers the Coffee Board officers to enter the premises of the estates, conduct inspections and take legal action against the growers if the estates were found negligent. However, the Coffee Board claims that Section 26 was intended to control the spread of White Stem Borer (WSB), Coffee Leaf Rust (CLR), Coffee Berry Borer (CBB) and other epidemics in the coffee-growing regions. On the other hand, growers fear that this could further curtail their rights on their own land. The bill, however, does not mention the designation of the officers who were given such discreet powers to inspect and penalize the growers, nor does it specify under what conditions they could enter the private lands of the growers. This has created confusion in the minds of the growers.

The new Bill had also proposed to appoint two Lok Sabha and one Rajya Sabha member to be the members of the Coffee Board. However, the growers presume that these politically elected representatives were likely to have less knowledge of coffee production aspects. As these members were appointed by the Union Government, there were also accusations from the growers that the Coffee Board would probably be controlled by the large growers, corporate, industrialists and politicians.

The growers were also apprehensive about the marketing aspects of the New Bill 2016, presuming that the marketing of coffee would transform from free trade to a restricted market with a rigid and authoritarian system. The organizations also presume that the growers would be mandated the requirement of TP3 form (Temporary Permit in form 3) from the Central Excise Department, which was prevalent prior to liberalization. According to this rule, the growers were permitted to carry coffee only in vehicles that were allowed by the Excise Department (Joy, 2004).

To conclude, the proposed rationale of the Coffee Bill, 2016 was still unclear to the growers. There were general presumptions among the growers that the New Bill 2016, rather than protecting the interests of the growers, would invite restrictions in the existing coffee system. They also viewed that the New Bill 2016 would only increase the role of the Coffee Board while not benefiting the growers in any way. There was also increased fear among the growers and organizations that such regulations (registration and inspections) would result in the exploitation of small growers in Kodagu. In this regard, the institutions were viewed as constraints and rules of games by the growers.

8.4 Role of Coffee Board as an Institution prior to liberalization

The great economic depression in the 1930s, followed by the Second World War in 1939, had resulted in the loss of major foreign markets for coffee, where the exports of Indian coffee fell drastically from 9240 tonnes in 1938–39 to merely 1260 tonnes in 1944–45. The above consequences put the Indian coffee industry on the brink of collapse. In order to safeguard the industry from the crisis during this period, the erstwhile British Government in India established the Indian Coffee Market Expansion Board on December 21, 1940, with the central objective of controlling, marketing and promoting Indian coffee. This eventually led to the establishment of the Coffee Board in 1942. With the establishment of the Coffee Board, the Indian coffee industry witnessed a period of recovery and stability.

Realizing the fall in coffee exports during the 1930s, the focus of the Coffee Board was diverted to the promotion of domestic markets. For the same reason promotional activities on domestic coffee consumption were undertaken by the Board during the mid-1940s, viz. running demonstration vans, incentivizing people on coffee consumption, sales of coffee powder and liquid coffee at subsidized rates, advertisements on coffee, establishment of the coffee houses, training on roasting, grinding and preparation of coffee and the establishment of a full-fledged propaganda department. By 1948, the Coffee Board handled 49 propaganda units that comprised Indian Coffee Houses (43), Coffee Bars (2), a Coffee van (1) and 3 coffee sales depots (Nanda, 1969).

During this period, the entire marketing of Indian coffee came under the control of the Board through the centralized 'pooled' marketing system. Here, the producers delivered the coffee (either dry or parchment) to the curing factories licensed by the Board, wherein the growers were paid well in advance for their produce. At the curing factories, processing operations (hulling, cleaning, sorting, grading and polishing) were undertaken. The costs of which were later deducted from the growers' payment. Further, the Coffee Board marketed coffee through two separate auctions, viz. 70 per cent

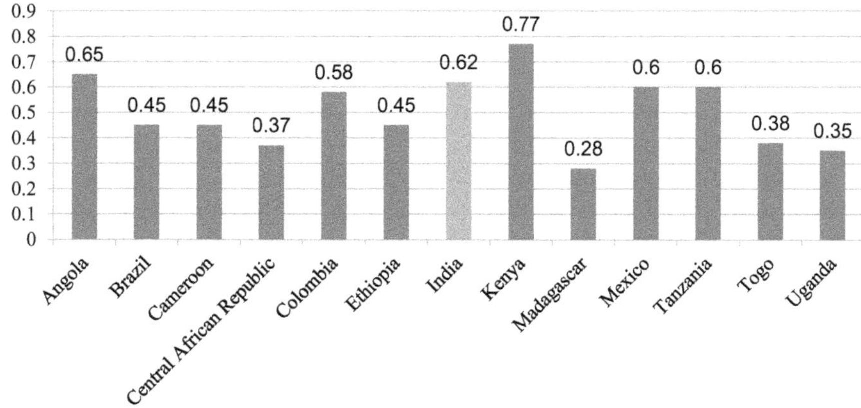

Figure 8.7 The ratio of producer price to ICO indicator prices prior to liberalization.
Source: Retrieved from Ekaterina (2007).

for exports and the remaining 30 per cent for domestic markets, although a major proportion of coffee was marketed through 'bulk contracts' of the Coffee Board.

The coffee producers were content with the centralized pooling system from the 1970s till the late 1980s. However, the decline in coffee prices, followed by the collapse of ICA negotiations in 1989 and the disintegration of the USSR which was a major importer of Indian coffee, had resulted in a decline in Indian coffee prices. The urge for liberalization was from the side of producers for the reasons mentioned below. The coffee market reforms are provided in Table 8.1a.

Firstly, the growers received prices much lower than ICO indicator prices. For instance, the Indian growers received merely 60 per cent of ICO indicator prices prior to liberalization (Figure 8.7). Secondly, the growers were dissatisfied with delayed payment (growers were paid in installments which varied between 1 and 3 years) for which the efficiency of the Board was highly doubted. Thirdly, the growers learnt that they were paying a substantial amount of sales tax, though coffee being an exportable commodity was exempt from paying sales tax.

8.5 Role of institutions in control of major pests and diseases[6]

8.5.1 Coffee Leaf Rust (CLR) and role of institutions

The Coffee Leaf Rust (CLR), a devastating disease of coffee plantations caused by the fungus *Hemileia vastatrix*, was first recorded in India in 1869, likely spread from Sri Lanka. The rust impairs the photosynthetic capacity of the leaf, resulting in defoliation and reduced initiation of flowers (Brown et al.., 1995), which could cause economic damage of up to 70 per cent in Indian Arabica (Prakash et al., 2004), while the severe infestation of leaf rust causes loss of foliage up to 50 per cent and berries loss up to 70 per cent (Bhat et al., 2000). Under normal conditions, the rust development comprises four phases, viz. the period of extension (April–May), period of intensification (Sep–Nov), period of defoliation (Dec–Jan) and period of inactivity (Jan–Mar), as indicated in Figure 8.8.

The pioneering works on this aspect were conducted by Wilson Mayne at CCRI in 1929. The findings showed that a noticeable breakdown of resistance to rust in coffee was effected by the emergence of new races of *H. vastatrix*, having virulence genes that could overcome the resistance genes of Coorg coffee (McCook, 2006). These fundamental works on rust were conducted in India, which further contributed to the establishment of the Coffee Rusts Research Centre at Oerias, Portugal, in 1955.

However, another measure to control CLR was quarantine by-laws operative under the Destructive Insects and Pests Act of 1914. According to this, the import permits assert the treatment of imported seeds with systemic fungicides, while the quarantine authorities checked the seed packages to inspect the quarantine regulations. Further, the Central Coffee Research Institute (CCRI) worked with the Protection of Plant Varieties and Farmers Rights Authority (PPV&FRA) to establish Distinctiveness Uniformity and Stability (DUS) for protecting and raising awareness on Intellectual Property (IP)-related issues in coffee. Meanwhile, coffee beans, freshly harvested berries, grafts, bud woods, seedlings or rooted cuttings from Africa, South America and Sri Lanka were strictly prohibited.

The more recent work of the Central Coffee Research Institute (CCRI), Balehonnur was the introduction of the 'Chandragiri' variety of coffee on 28 December 2007 with its original source of plant material from Portugal

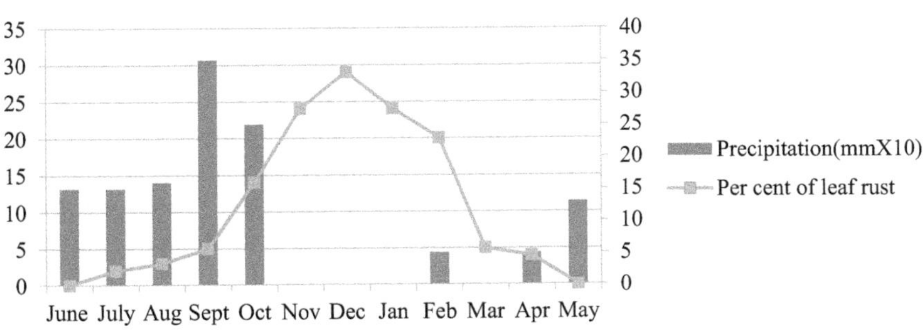

Figure 8.8 Coffee leaf rust development seasons.

Source: Own formulation.

from the hybrid which was a descendent of Villa Sarchi (semi-dwarf mutant of Bourbon coffee) and Hibrido de Timor. This variety of Arabica was found to be highly resistant to CLR (5–7 per cent), against the other varieties of Arabica (20–40 per cent).

In response to the worsening CLR, an action plan was carried out under the project "Increasing the resilience of coffee production to leaf rust and other diseases in India and four African countries" from 2008 to 2013. The project established Chandragiri and other Arabica varieties (Sln.5A, Sln.5B, Sln.6 and Sln.9) seed blocks at Coffee Board's research centres, wherein about 13,395 kg of Chandragiri seeds were produced and about 11,630 kg of other Arabica seeds (Sln.5A, Sln.5B, Sln.6 and Sln.9) were produced. Around 12,689 and 3244 beneficiaries from Karnataka and Tamil Nadu were provided with CLR-resistant seeds.

Although Chandragiri was found to be resistant to CLR, however, our field survey revealed that only 20 per cent of Arabica growers cultivated this variety, of which 25 per cent were large, 22 per cent were small, 15 per cent were medium and 12.50 per cent were marginal growers (Table 8.2a). The probable reasons could be that Chandragiri was a relatively new variety that was released only in 2007. Secondly, Chandragiri was more suitable at higher altitudes above 1015 metres above mean sea level (Narayana, 2014).

8.5.2 Coffee Berry Borer and role of institutions

The Coffee Berry Borer (*Hypothenemus hampei*) is another most dreaded pest said to be a native of Central Africa (Waterhouse and Norris, 1987). Probably, the Coffee Berry Borer entered into India accidentally from Sri Lankan refugees or through the illegal importation of coffee seeds (Singh and Ballal, 1991). In India, the berry borer infestation was first noticed in Gudalur region of the Nilgiri district of Tamil Nadu in February 1990 that later spread to nearby Coonor taluk of Nigiris and then later on to the neighbouring state Kerala, viz. Sultan Bathery and Vythiri taluks of Wayanad.

The berry borer infestation was first recorded in Karnataka in the Kutta region of Kodagu district during February 1991. By 1993, the Siddapur region of Kodagu was severely infested by this pest. By October 1999, the pest was also identified in the Hassan district of Karnataka (Bentley and Baker, 2002). The spread of the pest was so rapid that by the end of March 2001, the berry borer attack was recorded in all the three traditional coffee-growing regions of India. The infested area accounted for 35 per cent of the total area under coffee (Figure 8.9), while the maximum incidence of berry-borer infestation occurred in the month of September (Figure 8.10).

The Coffee Berry Borer (*Hypothenemus hampei*) being a new entrant to India was a big challenge to both growers and scientists as well. Thus, the first major joint survey on this pest was conducted in September 1990 by the four major institutes viz., Coffee Board of India, Directorate of Plant

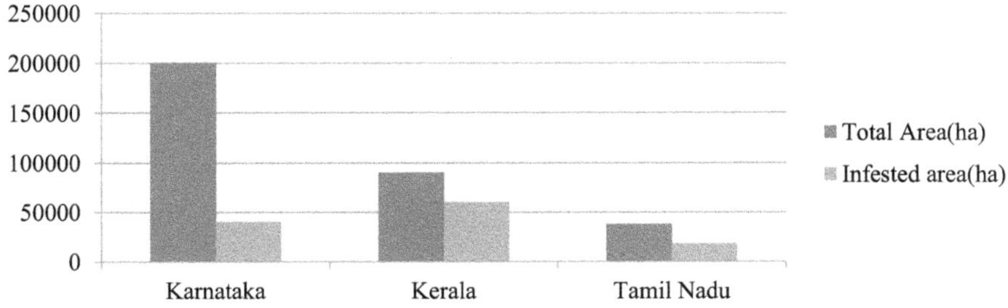

Figure 8.9 Infested area under coffee as of March 2001.

Source: Bentley and Baker, 2002.

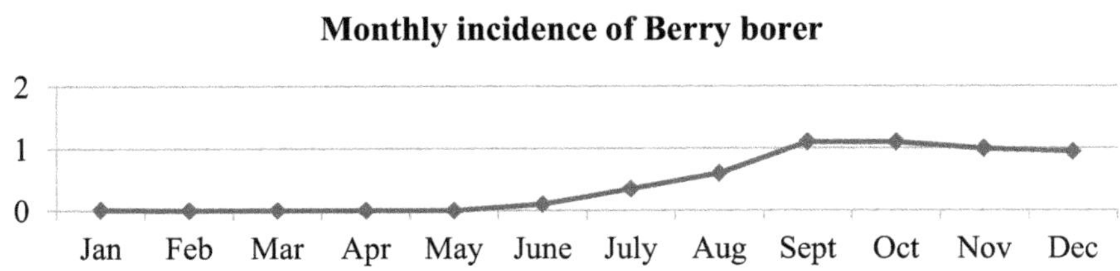

Figure 8.10 Monthly incidence of Coffee Berry Borer.

Source: Own formulation.

Protection, Quarantine and Storage, UPASI (United Planters' Association of South India) and Directorates of Horticulture (Government of Karnataka and Tamil Nadu). The coffee estates and curing works in South India were screened for plant and berry borer infestations.

Secondly, the Coffee Board launched a programme wherein the small growers were provided pesticides and plant protection chemicals at the subsidized rates thereby persuading the growers that the berry borer could be kept under control if timely pesticide sprays were undertaken. Besides the Coffee Board Extension department conducted innumerable field visits, workshops, seminars, camps and campaigns in creating awareness and motivating the growers to take up control measures against Coffee Berry Borer infestation, as shown in the Table 8.3a.

The Coffee Berry Borer had been notified under section 4A of the Destructive Insects and Pests (DIP) Act on 18 November 1992, wherein legal restrictions were imposed on the movement of coffee seeds, plants and coffee powder from Kodagu district (Karnataka), Nilgiri district (Tamil Nadu) and Wayanad district of Kerala to prevent the invasion of coffee berry borer to non-infested areas. The Coffee Berry Borer was covered under Domestic

Quarantine Regulations and the coffee growers were also directed to process their coffee only in identified curing works.

During 1994 the Scientific Advisory Committee chairman Dr M.S. Swaminathan put forward to consider the Coffee Berry Borer (*Hypothenemus hampei*) as a national pest and to execute a mission mode project in an effective manner. Following this, in December 1995 the two parasitoids namely *Cephalonomia stephanoderis* and *Prorops nasuta* were imported from Mexico and field releases were made in early 1996 in several infested areas of Kodagu. During this period, the growers were also advised on the phytosanitary measures (timely harvest, complete gleaning, removal of left-overs and strict drying standards) and cultural measures (pruning of coffee bushes to reasonable sizes, removal of coffee trees through the boundaries and judicious pesticide application).

Meanwhile, two research projects were sanctioned during December 1995 and January 1998. The former project was the Government of India-sanctioned project 'National Mission on Control and Prevention of Coffee Berry Borer', which emphasized educating growers on berry borer management, and the latter project was undertaken by the Common Fund for Commodities (CFC) and International Coffee Organization (ICO), viz. CFC-ICO project 'Integrated Management of Coffee Berry Borer' that facilitated training to the entomologists on Coffee Berry Borer.

Further various types of traps were field experimented and were well received by the growers who were encouraged to install the traps around the drying yards. The installed berry borer traps baited with ethanol–methanol and coffee fruit extracts were found to be most effective in borer trapping. In order to understand the effectiveness of the traps, a preliminary study was conducted in the Wayanad district of Kerala in both the harvested and non-harvested fields as shown in Table 8.4a. The multi-funnel traps were found to be highly effective that trapped about 668 and 72 borers in harvested and non-harvested fields per day, respectively.

8.6 Role of institutions in sustainable coffee production

'Sustainable coffee' refers to the coffee cultivated under a specific set of sustainable standards (social, economic and environmental standards), accredited and verified by independent third-party certification organizations. If the production practices were in compliance with the standards, then a certificate was issued to the producers. The certified coffee producers were then authorized to use 'logos' or 'trademarks' to market their products, which fetched a higher price in the market, besides fulfilling consumers' trust and taste. In this context, sustainable standards could be considered as 'internal market institutions' that ensured a higher quality of coffee from producers to consumers. The major certification schemes for coffee are provided in Table 8.5a.

In order to improve the quality of coffee, the Coffee Board encouraged the growers to take up eco-certification programmes, wherein subsidies were provided to the individual growers owing up to 20 ha, while the growers/ small growers' groups/collectives were encouraged to take up certification on various sustainability standards (Utz, Bird-friendly, Rainforest Alliance, organic and fair trade).

In the case of growers claiming for organic certification, individuals or beneficiaries were to be certified by the National Program for Organic Production (NPOP) even though they were also certified for other international standards, viz. European Union (EU), Japanese Agriculture Standards (JAS) and National Organic Programme (NOP) of USA. Here, the individual grower/ collectives were given a subsidy at 50 per cent of the certification costs with a ceiling of ₹50,000 per individual or group, wherein the amount would spread over for more than one year that also included the costs incurred during the conversion period (usually 3 years). Likewise, in case of other eco-certification programmes (Utz, Bird-friendly, Rainforest Alliance, organic and fair trade), the individual beneficiaries were eligible for 50 per cent of certification costs (maximum of ₹50,000) for one year only, whereas the support would be extended over more than a year for a group of small growers/collectives.

Besides, the Coffee Board also provided technical support to the coffee growers to adopt appropriate production practices, thereby facilitating them to take up sustainable eco-certification initiatives. For instance, a capacity building programme on 'Certified coffee for profitability and sustainability in the coffee sectors' was conducted for the coffee growers and Self-Help Groups in coffee-growing regions in collaboration with Indian Institute of Plantation Management (IIPM) in several coffee-growing regions. The capacity-building programme was conducted mainly to create awareness about certification programmes among small growers.

In order to promote organic coffee production in India, the Coffee Board had initiated the project 'Promotion of organic coffee production in India' during the IX plan period (1997–2002), which aimed at standardizing package of practices and creating organic awareness through training, workshops, seminars and educational campaigns, wherein initially financial assistance for the cost of inspection and certification was provided by the Coffee Board. The project was further extended to the X plan period (2002–07).

The Coffee Board is one of the major partners of CAFNET, a multi-country project that was initiated in 2007 aimed at connecting, enhancing and sustaining environmental services and market value of a coffee agroforestry system, which included even the small growers (prior to these only large industrial estates were certified). Nevertheless, the CAFNET along with its partners took initiative to incentivize the coffee growers in Kodagu to take up eco-certification (Utz, Bird-friendly and RFA) by organizing several workshops, capacity building programs and dissemination of information on sustainable coffee. The project was also closely associated with two leading coffee-trading companies (ECOM-Gill and Ned-Commodities) to promote eco-certification. The project was successful in certifying about 89 coffee growers and about 8 growers' groups for Utz and Rainforest Alliance certification in Kodagu district for which they received a premium that ranged between 5 and 10 per cent above the usual farm gate price during 2009–10.

The companies as a part of Corporate Social Responsibility (CSR) support the growers for sustainable practices through eco-certification. For instance, NESCAFE Plan, a global plan by Nestle company, was implemented in 2012 with an objective to improve the production practices of coffee so as to improve quality and sustainability in the long run. For this purpose, a coffee demonstration farm was set up near Sunticoppa town at Kodagu, Mudigere (Chikmagalur) and Kalpetta. It encouraged growers in obtaining certificate from the Common Codes for Coffee Community (4C). Besides, it conducted regular training to coffee growers on improving quality and technical assistance on eco-friendly practices.

The major problem in certification, particularly small growers, was the non-affordability of certification costs. Not all sustainably produced coffee is sold under a 'sustainable' tag and the growers thereby end up selling the coffee at the farm gate price itself. Additionally, the multitude of several certification schemes has also created chaos among growers, and other problems include the difficulty of record maintenance, which appeared to be cumbersome for growers Nonetheless, the lack of proper information on eco-certification had also left growers ignorant about the certification process. These reasons were quoted to be major ones for non-adoption of certification programmes among coffee growers.

The co-operatives in Kodagu remained inactive in encouraging the growers for the adoption of certification programmes. An active involvement of co-operatives could also prove beneficial for the adoption of sustainable coffee practices. One such instance was the success story of the Girijan Cooperative Corporation (GCC) in Andhra Pradesh for promoting 'Araku valley coffee' from the tribal tracts of Vishakhapatnam. Here the tribals were encouraged to cultivate organically grown coffee and were marketed by GCC at the National Commodity and Derivatives Exchange Limited (NCDEX) markets, thereby eliminating the middlemen in the process.

8.7 Expansion of area under coffee in non-traditional regions: promotion of Araku Valley coffee

Araku Valley is located in the Eastern Ghats (Vishakhapatnam and East Godavari districts of Andhra Pradesh), a non-traditional coffee-growing region and home to a number of tribal communities, wherein the semi-waste lands were converted into organic cultivation. Since 2000, the coffee has been introduced to the tribes in this region by the Integrated Tribal Development Agency (ITDA) supported by the Coffee Board with the assistance of non-governmental organizations, viz. Naandi Foundation and SKAL (now known as Control Union) to inspect and certify organic plantations. The coffee grown in this area was double certified with organic and fair trade and exported to countries like Norway, France and the United States (about 500 metric tonnes per annum).

To increase the area under coffee cultivation, the Coffee Board explored new coffee-growing potential regions such as Himachal Pradesh, Uttarakhand and West Bengal. The Board had initiated commercial cultivation of coffee in Bilaspur, Hamirpur, Mandi, Una, Kangra, Solan and Sirmaur districts of Himachal Pradesh, which resembled the foothills of Mullayagiri range of Chikmagalur. The temperature of Himachal Pradesh was best known to suit the Chandragiri variety of Arabica. The Central Coffee Research Institute (CCRI), Chikmagalur, has assured to provide coffee seedlings to this region for the cultivation of coffee.

8.8 Application of information technology (IT) in coffee production

The Kaapi Soil Health Management and Monitoring (Kshemam) app was launched by the Indian Institute of Information Technology and Management-Kerala (IITM-K) in order to provide necessary information on soil nutrient management and fertilizer application for their estates. The growers by filling in information on the geographical locations of the estates could also check for the nutritional status of soil and generate soil health cards in English, Kannada, Malayalam and Tamil languages. The National Bureau of Soil and Land Use Planning (NBSSLUP) has conducted GPS-enabled soil sampling and nutrient analysis in the three states to furnish growers with correct information.

Another innovative initiative by the Coffee Board was the mobile-enabled extension services provided to the growers through the pilot project 'CAFE MOVEL'[7] launched in 2013. The services were based on the Interactive Voice Response Technology (IVR). The technology was designed by the Centre for Agriculture and Bioscience International (CABI), also supported by the International Coffee Organization (ICO) and Common Fund for Commodities (CFC). The major functionalities included interactive learning on production practices, which alerted the growers on pests and diseases, weather conditions and prevalent market prices, while the grower is also subject to one-to-one interaction with the experts. Here, the helpline services of 24 × 7 are provided through the IVR, where the growers could record their queries with experts. However, useful analytics was also provided to various interest groups through web-based applications.

8.9 Major growers' associations in the coffee sector

The United Planters' Association of Southern India (UPASI) is an apex body of planters in Karnataka, Kerala and Tamil Nadu established in 1893. The association covered economic research, scientific research and publications, information exchange, market intelligence, industrial and public relations, training and estate supplies. UPASI represented growers' interest in national and state policies, viz. for setting facilities for clonal propagations and tissue culture planting, sustainable coffee production, associated lower productivity, import high yielding varieties of Arabica resistant to pests and diseases and exemption from GST up to the curing stage. Additionally it provides analytical services for soil amendments, fertilizers, manure, fungicide and pesticide applications, advises growers on the cultivation practices, organizes short-term industrial training programmes at the regional centres and undertakes research projects.

Karnataka Planters' Association (KPA) is an apex decision-making body in the plantation sector (coffee, tea, pepper, cardamom and rubber) established in 1958. The KPA makes representation to the concerned ministries on various issues faced by the industry, viz. environment and water management in coffee plantations, labour acts (wages and social costs), girl child corpus, drought in plantation regions, climate change and White Stem Borer in Arabica, man–animal conflict, decline in productivity, steep rise in costs of production, repeal of Coffee Act of 1942, collaboration to acquire high-yielding varieties, disbursal of subsidies, soil and leaf laboratories, establishment of the coffee kiosks and coffee park and exemption of Rule 7B (1) of Income Tax rules, to hold international coffee trade fairs at Madikeri and Chikmagalur.

Like the KPA at the state level, Codagu Planters' Association (CPA) is an apex district-level organization of growers that was established in 1879. CPA is also an active member of KPA and UPASI. The association promotes and safeguards the plantation industry of Kodagu, besides considering all the issues, represents policy decisions, analyzes and publishes statistics related to plantations and corresponds with Planter's and Employer's Association, trade unions and the government.

8.10 Role of institutions in marketing and exports of coffee

The Coffee Futures Exchange of India (COFEI) as a single commodity exchange was established in Bengaluru on December 22, 1997, in order to facilitate Coffee Futures trading under the Forward Markets Commission (FMC). The COFEI objectives comprised price discovery and standardized forward contracts. Since 1998, the COFEI has operated as a system of warehouse receipts. However, COFEI eventually proved to be infeasible and came to a halt in 2002.

The National Multi-Commodity Exchange (NMCE) in February 2005 with its headquarters in Ahmadabad began future trading in coffee. The same year it came to a halt due to speculative trading and issues related to contracts. However, NMCE re-launched future trading of Robusta coffee in Robusta estate pounded coffee, with 11.5 per cent of coffee contracts. However, since 2009 the NMCE's market share has been diminishing (1.3 per cent). The NMCE was also associated with Geojit Financial Services for online trading.

The National Commodity and Derivative Exchange Limited (NCDEX) headquartered at Mumbai was established in 2003. It was an electronic trading platform for trading agricultural and non-agricultural commodities that also included trading of coffee. The NCDEX comprised four promoter shareholders, viz. ICICI Bank Ltd, LIC, NABARD and NSE, while the other major shareholders included Canara Bank, PNB, CRISIL, IFFCO and regulated by the Securities Exchange Board of India (SEBI). The NCDEX along with the Coffee Board plan to establish an online platform for trading of coffee wherein the traders and growers make direct buy or sell decisions and effective price risk management such that growers realize 95 per cent of export prices. However, the COFEI, NMCE and NCDEX lacked interest from the traders and speculators due to poor speculative activity and contract specifications.

Multi-Commodity Exchange (MCX) was established in November 2003 and was based at Mumbai that commenced Robusta future trading from January 2007. The MCX was perceived to be the largest commodity exchange in the country with a market share of 72 per cent. The MCX worked under the regulatory framework of the Securities and Exchange Board of India (SEBI) and was promoted by NABARD, SBI, NSE, HDFC, Fid Fund (Mauritius) Ltd, Corporation Bank, Union Bank of India, Canara Bank, Bank of India and Bank of Baroda. The MCX had also undertaken bi-monthly contracts in the Robusta for instant coffee trade and entered into an agreement with Euronext life, which traded at a premium to LIFFE prices that notably boosted user confidence in the system. The MCX in collaboration with the Coffee Board established 13 warehouses in Kalpetta, Kerala and Chikmagalur.

8.10.1 Online trading initiatives in coffee

In addition to the above-mentioned coffee futures trading initiatives, online trading initiatives emerged in the coffee system. One such initiative was a web-based initiative by ITC Ltd, which established Coffee e-choupals in 2000. This initiative was undertaken to directly link the coffee growers with the company for the procurement of coffee. By the end of 2004, the company was successful in establishing 75 kiosks in 125 villages that approximately covered about 6000 growers. The company had launched a portal plantersnet.com for coffee growers in Karnataka to provide production, marketing information and advice to the growers. Subsequently, it included 'tradersnet .com', an online market for the purchase and sales of raw and clean coffee. This provided a platform for several stakeholders, which also had a tie-up with the world's largest future player ABN AMRO that provided information on coffee futures on a regular basis and the latest prices posted on commodity exchanges, viz. London International Financial Futures and Options Exchange (LIFFE) in London and Coffee, Sugar & Cocoa Exchange (CSCE) in New-York. The ITC is one of the major exporters of Indian coffee-initiated community development programmes in non-traditional coffee-growing regions, viz. Orissa and Andhra Pradesh, the non-traditional coffee-growing regions. The ITC assisted the growers to cultivate pesticide-free coffee under the guidance of the Coffee Board, besides marketing the end product effectively.

In the case of online auctions, the sellers submit the necessary information on the product, which mentions the variety, grades, quantity, price and details of the seller. The online auctions were held for 5 days a week between 9 a.m. to 11 a.m., through ascending English auctions. In this case, the bids were announced, openly viewable and open for around 3 minutes after which the product sold out to the highest bidder. The unsold lots were again submitted for the bidding process on the following days or marketed through other channels. The major advantages of auctions were lower transaction costs.

8.11 Major associations involved in coffee processing and exports

The coffee traders formed an association known as the Indian Coffee Traders Association (ICTA) in February 1993 to conduct coffee auctions (weekly auctions on every Thursdays) in the premises of the Coffee Board on behalf of the growers or sellers other than conducting auctions in other coffee-growing regions. The growers/sellers exhibit the cured coffee (Green beans), certified specialty coffee and organic coffee for which the traders, roasters and exporters bid the coffee, and the highest bidder is offered the sample if the price meets the auctioneer's approval. The auctioneers charge a fee of 1 per cent from the sellers as brokerage value. It was estimated that approximately about 600 tonnes of coffee was offered for auction every week.

The Specialty Coffee Association of India (SCAI) is a non-profit organization aimed to expand a sustainable specialty coffee trade through education, research, events and carved a niche for Indian coffee by promoting sustainable, organic, geographic and estate branded coffee. The members comprised growers, traders, roasters, retailers, brokers, curers, auctioneers and input suppliers and stakeholders who seek to promote specialty coffee. The SCA holds the India Barista Championship (IBC) and Coffee Mela wherein the growers were encouraged to display different coffee varieties, create awareness of best plantation practices, design right blends and preparation of espresso-based coffee drinks. The winners of IBC were selected to participate in the annual World Barista Championship (WBC). The SCAI plans for the digital auctions of certified specialty coffee and promote cafe culture through baristas.

The All-India Coffee Curers Association (AICCA) was the only representative body of the coffee curers. The Association engages in facilitating processing and engages in policy dialogues pertaining to curers and represents curers' interests in various forms, viz. demands for a revival of closed curing centres, withdrawal of 18 per cent GST on coffee curing, modern curing machinery, recommend government for financial support for up-gradation of curing industry, processing at curing works to be treated as agricultural produce, labour negotiations and allowing retaining allowances to labourers, renewal of licenses for pollution control, negotiation of curing charges and looks into the quality issues such as sieve size grading.

The Indian Coffee Roasters Association, which represents the interest of coffee roasters, has been promoting differentiation strategy and high-quality segment of markets on the basis of blends and quality grades; it promotes specialty coffee roasters, high-tech roasting rather than traditional pan-frying method, value addition, the development of local roasting capacity, train roasters and soluble manufacturing capacities. However, the roasters and traders' associations lack collective memberships and are fairly weak.

Likewise, the Coffee Exporters Association for whom the membership was based on selective basis, with concrete knowledge of international coffee laws and regulations. The Association furnishes members with commercially useful market information, market network, organizes trade fairs, access to special markets and technical advice.

8.12 Role of private institutions in coffee retailing

The Cafe Coffee Day (CCD) is a division of Amalgamated Bean Coffee Trading Company Limited (ABCTCL). The first Cafe Coffee Day (CCD) is the largest retailing coffee outlet in India, launched on July 11, 1996, in Bangalore. The coffee was fetched from 10,000 acres of Arabica plantations in Chikmagalur. The CCD identified and encouraged prospective of Cafe culture in India. By this the CCD promoted domestic consumption of coffee especially among the youth and recognized as the most popular hangout joint among the youth. The offshoots included over 2000 Cafe Coffee Day joints, around 400 Fresh & Ground stores in the country and around 1000 Coffee Day Xpress kiosks, with the logo being, "*A lot can happen over coffee*". The CCD also sells merchandise like T Shirts, bags, coffee filters, coffee powders and coffee mints.

The Barista launched its first outlet in Delhi in 2000, which is the second-largest operator of coffee outlet in India. This was taken over by Lavazza an Italian company in 2007 and known to brew Italian coffee. By 2012 Barista was established in about 30 cities in India with over 200 Barista Lavazza Espresso bars. The Barista mainly source coffee from TATA coffee. While selling merchandize has been a part of coffee retailing in Barista, viz. coffee mugs, Blue Curacao, Barista French Press and Barista coffee beans which are mostly imported. The Barista has tie-ups with Taj Group of hotels, Planet M and corners. Another strategy includes setting up Bancafe along with ABN AMRO, viz. cafe in the bank premises.

Qwiky, the first coffee pub in India was launched at Chennai in October 1999 with Chimayo Chains as a master franchise for Qwiky's coffee pubs. There are about 22 outlets in India, and they have also initiated Shop-in Shop (SIS) gourmet coffee bars. These coffee kiosks have been set up at the Bharat Petroleum pumps and advanced 12 coffee islands in retail stores such as Lifestyle, Ebony, Pyramid and Music World.

The Starbucks, an American coffee chain, entered as a joint venture (50:50 joint ventures) with Tata Coffee, popularly known as Tata Starbucks. The coffee was mainly sourced from TATA coffee estates. The first outlet in India was set up in 2012 in Mumbai. Currently, the Starbucks have more than 50 outlets across Delhi, Mumbai, Pune, Bengaluru and Chennai. The Indianization of ambience of store and food menu were the main reasons for success of Starbucks in India. The other private players in the coffee sector includes Gloria Jeans, Costa Coffee, The Coffee Bean & Tea Leaf, Di Bella and so on.

8.13 Role of Coffee Board in exports of coffee

After liberalization the exports of coffee was being carried out by the private exporters, while the Coffee Board acted as a facilitator and promoter of Indian coffee exports and domestic consumption. One of the initiatives to promote domestic consumption was to launch coffee stamps, wherein the GOI and Coffee Board of India in partnership with India post launched the coffee flavoured miniature sheet stamp on 23 April, 2017, in Bengaluru (printed at the India Security Press, Nasik and Maharashtra). The aromatic coffee stamp was a way to promote Indian coffee by sensitizing people to taste coffee. The stamp had in it the pictures of coffee beans and a cup of brew and word 'COFFEE' in it, being sprayed with the fine granules of coffee to retain its aroma.

The Kaapi Shastra programme was a short-term five-day training programme organized by the Coffee Board to encourage domestic consumption, which also aimed at improving quality of coffee through value addition. The programme created awareness on coffee roasting and brewing activities meant mainly for the roasters, grinders, hotel/restaurant/cafeteria personnel, institutional caterers and entrepreneurs in the retailing sector for value creation, while also aimed to sustain the coffee industry by creating a strong domestic market for coffee.

However, the Board authorized the registration of the exporters. Currently, about 395 exporters were registered with the Coffee Board. Additionally, the Coffee Board as per section 20 of the Coffee Act, 1942, issues 'certificates of origin' so as to ensure that every export was covered under ICO certificate of origin. The Coffee Board also participates in the international centric/food and beverage exhibitions wherein the exportable grades of Indian coffee were promoted.

In order to encourage exports of Indian coffee and maximize their export performance, the Coffee Board awards the eight top most exporters of India. The best exporter of green coffee, specialty coffee and instant coffee are being awarded, while other five export awards were institutionalized according to the country of export based on the quantity exported, viz. the United States, European Union, Russia and Commonwealth of Independent States (CIS), Middle East and Far East regions. Another way for export promotion was to provide an incentive at ₹3 per kg of value added for producers' coffee exported as Indian brand, while an export incentive of ₹2 per kg to high value markets (Australia, Canada, Finland, Japan, New Zealand, Norway and the United States). The advertisements on the uniqueness and excellence of Indian coffee were published in the notable trade journals and magazines.

8.14 Major schemes for promotion of exports

Referring to the Foreign Trade Policy (2015–20) implemented under the Department of Commerce, Ministry of Commerce and Industry through Merchandise Exports from India Scheme (MEIS); previously, there were five distinct schemes, viz.. Focus Product Scheme, Market Linked Focus Product Scheme, Focus Market Scheme, Agri. Infrastructure Incentive Scrip and Vishesh Krishi and Gram Udyog Yojana (VKGUY). The Vishesh Krishi Gram Udyog Yojana (VKGUY) was a scheme in the foreign trade policy from the GOI to support the exports of agricultural products. Under this scheme, green coffees were eligible for incentive in the form of Duty Credit Scrip at 3 per cent of the Free on Board for coffee exports. However, now these five schemes were merged into MEIS as per the Indian Foreign Trade Policy 2015–20. The MEIS intended to counteract the infrastructural inefficiencies and the associated costs involved during the exports of coffee and other commodities with high export potential, while also enhancing the export competitiveness in the sector. Here the duty credit scrips[8] were provided under MEIS. Such incentives shall be utilized for the payment of Customs duties for import of inputs, excise duties payment on domestic procurement of goods, service and customs duties.

The Market Development Assistance (MDA) schemes were also provided for the exporters. The Department of Commerce assists the exporters for export promotion and to undertake non-recurring innovative activities. Another way for export promotion was to provide an incentive at ₹3 per kg of value added for producers' coffee exported as Indian brand. While an export incentive of ₹2 per kg for coffee exported to high value markets (Australia, Canada, Finland, Japan, New Zealand, Norway and the United States). Besides, the Coffee Board also participates in the coffee centric/food and beverage exhibitions abroad.

8.15 Summary

Firstly, the chapter views institutions as norms, rules and strategies that govern human activities (land tenure system in Kodagu), which had a moderate impact on the agro-forestry system and production systems of Kodagu. Secondly, it perceives institutions as legal and political regulation Coffee Bill, 2016, wherein Coffee Act of 1942, with minimal modifications had been drafted as the Coffee Act of 2016, which was much to the dissatisfaction of the growers, who presume the Coffee Bill, 2016 as humanly devised constraints. Thirdly, it considers institutions as a particular type of social structure (Coffee Board), its role prior to liberalization and the 'hard institutional failure' that paved way for liberalization.

The next section discusses the role of institutions in controlling major pests and diseases in India. Further, it discusses the role of institutions in sustainable production and marketing (organic, RFA, Utz, 4C and Nespresso) and the role of the Coffee Board, NGOs, SHGs and corporate. The expansion of area under coffee in the non-traditional regions, CAFE Models, coffee stamps and programmes and promotional activities for exporters, application of information technology, major growers' associations, institutions like COFEI, NMCE, NCDEX, MCX, online trading, association in coffee processing and exports, role of private institutions and major schemes for coffee exporters

Overall, it could be concluded that during the post-liberalization period, the institutions played a phenomenal role in getting distinct identity for Indian coffee in the global markets in terms of coffee production, marketing and exports. However, still institutions have failed to provide satisfactory remedies for most devastating pests like WSB; which was first identified in 1838. In this regard, a much aggressive and inter-continental, combined research is required.

Notes

1 Silver oak was an alternative source of income to coffee growers owing to its timber value. The growers did not require permission from the government for

cutting and selling of silver oaks. Moreover, pepper vines were crept on silver oak trees.

2 The Ministry of Commerce and Industry released the draft Coffee Bill 2016 on March 23, 2016.
3 While Section 16 mandates the registration of the curing establishments, Section 17 mandates the registration of every roasting, grinding and instant coffee manufacturing.
4 The Coffee Act of 1942 was established by the British government during the Second World War in order to keep the coffee industry under its control, while the growers were not permitted to market their coffee without the consent of the government. The Act also restricted the growers from transporting the produce as it required the permission of Central Excise Authorities even to transport from one place to another.
5 The differences between Coffee Act of 1942 and Draft Coffee Act, 2016, are highlighted in bold letters.
6 Role of institutions for control of WSB has been provided in Chapter 5.
7 CAFÉ MOVEL means mobile coffee in Portuguese language.
8 Duty credit scrip is the export promotion incentive provided by GOI.

References

Abraham, R., Purushothaman, S., and Devy, S., 2013, Conservation and coffee production: Creating synergies in Kodagu, Karnataka. In *Livelihood Strategies in Southern India: Conservation and poverty reduction in forest fringes*, pp. 89–107, Springer India.

Bentley, J.W. and Baker, P.S., 2002, *Manual for collaborative research with smallholder coffee farmers*. CABI Commodities, Egham, 130 pp.

Bhat, S.S., Naidu, R., Daivasikamani, S. and Nirmala Kannan, 2000, Integrated disease management in coffee. In Daivasikamani, S., RajanaikaUpadhyay, R.K., Mukerji, K.G. and Dubey, O.P. (eds.) *IPM system of agriculture-cash crops*. Volume 6, Aditya Books Private Limited, New Delhi, 564 p.

Brown, J.S., Whan, J.H., Kenny, M.K. and Merriman, P.R., 1995, The effect of coffee leaf rust on foliation and yield of coffee in Papua New Guinea. *Crop protection*, **14**(7): 589–592.

Businessline, 2017, Available at https://www.thehindubusinessline.com/economy/agri-business/isros-eye-in-the-sky-to-put-new-coffeegrowing-areas-on-the-map/article9640419.ece

Coase, R.H., 1937, The nature of the firm. *ECONOMICA*, **4**(16): 386–405.

Commons, J.R., 1951, *The economics of collective action*. Macmillan Company, New York.

Crawford, S. and Ostrom, E., 1995, A grammar of institutions. *The American Political Science Review*, **89**(3): 582–600.

Damodaran, A., 2002, Conflict of trade-facilitating environmental regulations with biodiversity concerns: The case of coffee-farming units in India. *World Development*, **30**(7): 1123–1135.

Ekaterina, K.G., 2007, *Three essays on coffee markets*. Doctoral dissertation. University of Maryland, College Park.

Garcia, C.A., Bhagwat, S.A., Ghazoul, J., Nath, C.D., Konerira, M.N., Cheppudira, K.G., Raghuramulu, Y., Nasi, R. and Vaast, P., 2010, Biodiversity conservation in agricultural landscapes: Challenges and opportunities of coffee agroforests in the Western Ghats, India. *Conservation Biology,* **24**(2): 479–488.

Hodgson Geoffrey, M., 2006, What are institutions? *Journal of Economic Issues,* **40**(1): 1–25.

Joy, C.V. 2004, *Small coffee Growers of Sulthan Bathery, Waynad.* Discussion Paper No. 83, Kerala Research Programme on Local Level Development, Centre for Development Studies, Thirunvanthapuram.

McCook, S.G., 2006, The global rust belt: Hemileia vastatrix and the ecological integration of world coffee production since 1850. *Journal of Global History,* **1**: 177–195.

.Nanda, A.C., 1969, Union response to crisis. *Indian Journal of Industrial Relations,* **4**: 508–522.

Narayana, M.R., 2014, Economic analysis of leaf rust management by chemical controls: Evidence and Implications for household coffee farmers in India. *Indian Journal of Agricultural Economics,* **69**(4): 1–15.

North, D.C., 1989, Institutions and economic growth: An historical introduction. *World Development,* **17**(9): 1319–1332.

Ostrom, E., 2010, Beyond markets and states: Polycentric governance of complex economic systems. *American Economic Review,* **100**(3): 641–672.

Prakash, N.S., Marques, D.V., Varzea V.M.P., Silva M.C., Combes M.C. and Lashermes, P., 2004, Introgression molecular analysis of a leaf rust resistance gene from Coffea liberica into Coffea arabica L. *Theoretical and Applied Genetics,* **109**(6): 1311–1317.

Sathish, B.N., Kushalappa, C.G. and Vachana, H.C., 2012, *Tree diversity in coffee plantations in different land tenure systems.* LAP Lambert Academic Publishing, Germany.

Singh, S.P. and Ballal, C.R., 1991, *Status of coffee berry borer* (*Hypothenemus hampei* (Ferr.) (Coleoptera: Scolytidae). Technical Document No. 36, AICRP on biological control of crop pests and weeds, Bangalore. p. 6.

Scott Tyler, A., 2015, Does collaboration make any difference? Linking collaborative governance to environmental outcomes. *Journal of Policy Analysis and Management,* **34**(3): 537–566.

Tenaw, S.K.M., Zahidul, I. and Parviainen, T., 2009, *Effects of land tenure and property rights on agricultural productivity in Ethiopia, Namibia and Bangladesh.* Discussion Papers No. 33. University of Helsinki Department of Economics and Management, Helsinki.

Tulu, Z.J., 2010, *Institutions, incentives and conflict in coffee forest use and conservation: The case of Yayo Forest in Iluu Abba Bora Zone, Southwest Ethiopia.* Doctoral dissertation. University of Bonn, Germany.

Uthappa, K.G., 2004, *Land tenure, land holding and tree rights of Kodagu.* Kodagu Model Forest Trust, Ponnampet.

Veblen, T., 1934, *The theory of the leisure class: An economic study of institutions.* Introduction by Stuart Chase. The Modern Library, New York.

Waterhouse, D.F. and Norris, K.R., 1987. *Biological control: Pacific Prospects.* Inkata Press, Melbourne, Australia, 454 pp.

Williamson, O.E., 1985, The economic institutions of capitalism: Firms, markets, relational contracting, Free Press/Collier Macmillan, New York/London.

Chapter 9

Summary and conclusions

9.1 Introduction

Coffee is an important commercial crop of the tropics, highly demanded all over the world and thereby very important for international trade. It occupies a significant position in the Indian economy, wherein the country earned a foreign exchange worth ₹5795.71 crore from the export of coffee in 2017–18. India has a 3.30 per cent share in the total world production and 5.02 per cent in total exports.

In India, Karnataka stands first in area, production and productivity. The area under coffee cultivation in Karnataka is 2.44 lakh hectares, accounting for 54.47 per cent of total country's area under coffee. The state contributes to 71.10 per cent of the total country's production. In Karnataka, the Kodagu district stands first in area and production of coffee, contributing 53.06 per cent of the state's production and 38.09 percent of country's production.

At a microlevel, several factors influence the productivity of coffee, either positively or negatively. Therefore, it is necessary to identify the major factors that influence (enhance or limit) the productivity of coffee in the study region. Additionally, the cost of cultivation is also increasing every year. With the rise in cost structure, does coffee cultivation remain profitable to the growers or are they incurring losses? Furthermore, producers, traders and exporters face several problems in maintaining quality. Considering the problematic situation of the coffee industry, it is necessary to investigate the production and marketing aspects of coffee in Kodagu. Therefore, this study was carried out with the following objectives:

1. To study the major factors influencing the productivity of coffee in Kodagu.
2. To study the economics of Arabica and Robusta varieties of coffee in the study region.
3. To study the existing marketing structure and pattern followed in coffee marketing and analyze problems faced by producers, traders and exporters in the maintenance of quality,

DOI: 10.4324/9781003541721-9

4. To study the role institutions played in the process of coffee production, marketing and export.

The present framework of the study was based on primary data collected from the selected sample of coffee growers from three Arabica (Somwarpet, Shanivarsanthe and Sunticoppa) and Robusta (Virajpet, Siddapura and Napoklu) liaison zones of Kodagu. The pre-tested questionnaires were used to collect the data from the growers, classified as small (<2 ha), semi-medium (2–10 ha), medium (10–25 ha) and large (>25 ha). The study comprised 120 growers (60 Arabica and 60 Robusta growers). Separate checklists were used for market functionaries. The data pertains to the reference period of the 2015–16 agricultural year. Multiple regression analysis was carried out to understand the factors influencing the productivity of coffee, factors affecting the incidence of White Stem Borer and net price of Arabica and Robusta growers; descriptive statistics, augmented Dickey–Fuller (ADF) unit root for stationarity tests, Johansen co-integration test and Granger Causality test were used for predicting price relationships. The study also used different cost concepts and tabular presentations.

9.2 Summary of findings

The findings of the study are given below in brief.

1. The productivity of coffee varied among different varieties. In the case of Arabica, the productivity of coffee among the sample growers ranged between 600 and 1200 kg/ha, the mean productivity being 803 kg/ha. In contrast, for Robusta, the mean productivity was 1590 kg/ha, with the lowest being 1250 kg/ha and the maximum being 1800 kg/ha.
2. The mean age of Arabica and Robusta trees was 28.02 and 30.58 years, respectively. The age of Arabica trees ranged between 15 and 46 years, while that of Robusta trees between 10 and 60 years. The productivity of coffee initially increased with age, reached its maximum and then decreased with age. Specifically, the productivity of Arabica and Robusta increased up to 24 years and 30 years, respectively, where the inflection point was observed. The age and productivity of coffee exhibited an inverse U-shaped relation. The results of regression analysis indicate that in both the varieties, the co-efficient associated with age was found to be positive, while the age square was negative.
3. The second factor considered was the elevation of estates. The elevation of estates ranged between 853 and 1026 m.a.s.l. in Arabica while that of Robusta ranged between 704 and 1072 m.a.s.l. The coefficient associated with elevation was found to be positive in Arabica (0.36). In contrast to this, the coefficient associated with Robusta was found to be negative (-0.16). In other words, the productivity of Robusta declined

with elevation, whereas Arabica productivity had a direct relation with elevation.

4. The mean annual rainfall in the surveyed estates ranged between 1397 mm and 2413 mm for Arabica, while the Robusta estates received heavier rainfall that ranged between 1058 mm and 4656 mm. The influence of average annual rainfall on the productivity of Arabica was found to be positive and vice versa in Robusta; however, this relationship turned out to be insignificant in the model. On the other hand, the critical rains had a positive influence on the productivity of Robusta.
5. The average number of shade trees was found to be higher in Arabica estates (237 shade trees per ha) compared to Robusta estates (147 shade trees per ha). The productivity of Arabica coffee increased with an increase in shade and was found to be significant at 1 per cent. However, the shade trees had a negative influence on the productivity of Robusta but turned out to be insignificant.
6. The White Stem Borer (WSB) associated with Arabica trees had a major negative influence on the productivity of Arabica, which turned out to be highly significant. The coefficient associated with the White Stem Borer was –7.66.
7. Among the surveyed Arabica growers, about 68.33 per cent of the growers were shifting towards Robusta, 21.67 per cent of the Arabica growers were willing to shift but unable to shift, while only about 10 per cent of the growers were not shifting towards Robusta. A five-point Likert scale (strongly agree received a ranking of 5, while strongly disagree received a ranking of 1) was used to understand the factors responsible for the shift, not shifting and willingness to, but unable to, shift among the Arabica growers.

   About 90 per cent of the growers who were shifting towards Robusta strongly agreed that the heavy infestation of White Stem Borer (*Xylotrechus quadripes*) and resultant losses in the crops had prompted them to shift towards Robusta. On average, about 150 trees per ha were uprooted in individual estates, the average yield loss being 5.45 per cent. The survey also revealed that the higher incidence of borer attack was observed in S.795 and Sln.6 varieties of Arabica.

   The Arabica variety of coffee was also highly susceptible to Coffee Leaf Rust (CLR) caused by *Hemileia vastatrix*. However, our survey found that currently CLR disease was under control due to the regular application of Bordeaux mixture. Likewise, about 90 per cent of the growers opined that CLR was not the cause for the shift. About 80 per cent of the growers moderately asserted that the labour requirement was higher in the case of Arabica and that the severe shortage of labour had paralyzed the Arabica plantations in Kodagu and that was also one of the reasons for the shift. Further, about 30 per cent of the Arabica growers strongly agreed that the higher cost of management in Arabica had prompted them for the shift, while about 60 per cent of the growers took a neutral stand in this regard, and about 10 per cent of the growers moderately agreed with the same.

About 21.67 per cent of the interviewed growers were willing to shift towards Robusta but were unable to do so. Of these, 80 per cent strongly agreed that the lack of irrigation constrained them from shifting towards Robusta owing its higher irrigation requirements, while about 20 per cent took a neutral stand on this. However, about 10 per cent of the interviewed growers were not willing to shift towards Robusta; of these, 80 per cent of the respondents strongly believed in the bright future prospects for Arabica and wanted to save Arabica for future generations.

8. The estates were classified in accordance with elevation (metres) and rainfall (mm). The estates (>1000 m.a.s.l.) and (<1000 m.a.s.l.) were classified as high elevation (HE) and low elevation (LE), respectively. Similarly, the estates receiving (1016–1778 mm) were classified as low rainfall (LR), those receiving (1778–2794 mm) as medium rainfall (MR) and those receiving (2794–50,580 mm) as high rainfall (HR). It was evident from the survey that the incidence of WSB was very high in estates located at low elevation and low rainfall (LE + LR), where 20 per cent of the estates suffered very high incidence, with the uprooting of trees ranging between 125 and 250 trees/ha. However, none of the surveyed estates were located in the high elevation and high rainfall belt (HE + HR). The incidences of the borer were relatively low in estates located in the high elevation and medium rainfall (HE + MR) belt.
9. The regression analysis on factors influencing the uprooting of Arabica revealed that the incidence of White Stem Borer decreased with an increase in altitude as the coefficient associated with elevation was found to be negative (–8.36). The number of shade trees per ha was found to be negative (–0.49), indicating that the uprooting of trees was lower in estates with a higher number of shade trees per ha. The incidence of WSB also decreased with an increase in expenditure on pesticides (–0.51) and was found to be significant at 5 per cent. However, the coefficient associated with the area under coffee and rainfall was found to be insignificant in the model. The study reveals that the incidence of WSB was higher within the age group between 20 and 30 years in Arabica, while the incidence was found to be lower in the coffee trees above 30 years of age.
10. The cultural methods included borer tracing, detection, hand-picking of beetles and uprooting and burning of infested plants. The costs varied between ₹670 and ₹7500 per ha of Arabica estates. The major chemicals were Chloropyrifos, Lindane and Contaf, along with combination of lime and neem. The chemical costs varied between ₹5393 and ₹6500 per ha, while the labour costs for chemical application varied between ₹2032 and ₹2500 per ha. About 56 per cent of the growers adopted

chloropyrifos, followed by Lindane (23 per cent) and chloro + lime by 9 per cent of the growers.

11. The institutional support included the distribution of pheromone traps by the Coffee Board at a subsidized rate, implementation of the Pest and Disease Act of 1968, wherein the borer was declared a serious pest, subsidy for gap filling in infested areas, demonstrations for popularizing IPM, field trials by CCRI and collaborative research by ICAR and BCRL.
12. The costs incurred in coffee were disintegrated into two components, viz. fixed costs (38 per cent and 36 per cent in Arabica and Robusta, respectively) and operational costs (62 per cent and 64 per cent in Arabica and Robusta, respectively). The total operational costs increased with an increase in size of the holdings, highest among the large growers (₹112,847 and ₹105,602 per ha in Arabica and Robusta, respectively), medium (₹107,964 and ₹93984 per ha in Arabica and Robusta, respectively), semi-medium (₹100,356 and ₹91712 per ha in Arabica and Robusta, respectively) and small (₹92,232 and ₹83,943 per ha in Arabica and Robusta, respectively).

The labour costs comprised 55 per cent of the total operational costs, which were further classified as the wages paid for the hired labourers and the imputed value of family labourers. The overall labour cost was found to be ₹56,839 and ₹50,655 per ha in Arabica and Robusta, respectively. In the case of Arabica, the total labour costs were the lowest among the small growers (₹53,109 per hectare), while in case of Robusta, it was lowest among the medium growers (₹49,476 per hectare). Overall, the hired labour was found to be ₹33,091 per ha in Arabica and ₹28827 per ha in Robusta. The family labour was found to be ₹23,748 and ₹21,828 per ha in Arabica and Robusta, respectively.

The input costs comprised manures, fertilizers, liming and dolomite, irrigation, weeding and pesticides. The overall input cost was ₹31,755 and ₹29,871 per ha in Arabica and Robusta, respectively. The manures and fertilizers accounted for the majority of total input costs, viz. 60 per cent and 65 per cent of total input costs. This was followed by irrigation costs that accounted for 22 per cent and 17 per cent of total input cost in Arabica and Robusta, respectively. The weeding accounted for 4.46 and 7.13 per cent in Arabica and Robusta, respectively (viz., ₹1418 and ₹2131 per ha in Arabica and Robusta, respectively). The overall processing costs in Arabica and Robusta comprised only 3 per cent and 2 per cent of the total costs, while the marketing costs in Arabica and Robusta were 2 per cent and 3 per cent of total costs, respectively.

The overall fixed cost in Arabica and Robusta was found to be ₹64,138 and ₹52,287 per ha. The rent of owned land accounted for 43 per cent and 59 per cent of the total fixed costs in Arabica and Robusta, respectively. The amortized establishment costs accounted for 29.45 per cent of the total fixed costs in Arabica and 17.65 per cent in Robusta, respectively. This was followed by interest on fixed capital, viz. 10.68 and 10.70 per cent of the total costs in Arabica and Robusta, respectively.

The total cost of cultivation increased with the size of holdings in both cases. It was found to be ₹142,280 per ha in small, semi-medium (₹162,441 per ha), medium (₹176,609 per ha) and large Arabica holdings (₹191,710 per ha). However, in Robusta, it was found to be ₹129,881 per ha in small, semi-medium (₹141,428 per ha), medium (₹150,181 per ha) and large (₹173,111 per ha). The overall cost of cultivation was found to be ₹166,697 and ₹144,811 per ha in Arabica and Robusta, respectively.

The net incomes for Arabica small and semi-medium were found to be ₹15,848 and ₹687, respectively. However, the medium and large growers incurred a loss of ₹13,109 and ₹23,376 per ha. The net income of small holdings was higher in the case of Arabica growers. In the case of Robusta, the net income was found to be ₹49,669 per ha among the small growers, followed by semi-medium (₹39,731 per ha), medium (₹42,041 per ha) and large (₹12,086 per ha). The overall net revenue was ₹39,852 per ha in Robusta.

The analysis of the cost of cultivation in Arabica shows that, on average, the total cost (Cost C_2) per hectare was highest among the large growers (₹191,710 per ha), followed by medium (₹176,609 per ha), semi-medium (₹162,441 per ha) and small (₹142,280 per ha). Likewise, in Robusta, this was found to be highest among the large growers (₹173,111 per ha), followed by medium (₹150,181 per ha), semi-medium (₹141,428 per ha) and small (₹129,881 per ha).

The cost of production (₹/quintal) in Arabica for small, semi-medium, medium and large growers were found to be ₹14,687 per quintal, ₹16,082 per quintal, ₹17,661 per quintal and ₹18,732 per quintal, respectively. However, it was found to be lower in Robusta viz., ₹4123, ₹4453, ₹4610 and ₹5452 per quintal, respectively among the small, semi-medium, medium and large growers.

In Arabica, the returns over variable costs were highest among the small growers (₹90,663 per ha), followed by semi-medium (₹64,507 per ha), medium (₹31,060 per ha) and large growers. (₹17623 per ha). Likewise, in Robusta, the returns over variable costs decreased with an increase in size of holdings, viz. small (₹119,938 per ha), semi-medium (₹102,653 per ha), medium (₹86,104 per ha) and large (₹53,033 per ha).

The family labour income was found to be negative among medium and large Arabica growers. However, in the case of Robusta, the family labour income in small, semi-medium, medium and large was found to be ₹85,281, ₹66,264, ₹49,232 and ₹15,798 per ha, respectively. The returns per rupee of investment were found to be higher among the small growers in both cases: for Arabica, small (1.11), semi-medium (1.00), medium (0.93) and large (0.88), whereas for Robusta, small growers (1.38), semi-medium (1.28), medium (1.27) and large (1.07).

In order to determine the revenue accrued per quantity of resource used in production, the gross income of the growers was regressed against labour use, fertilizers, irrigation, pesticides and weedicides. The labour use (man-days) in both varieties of coffee was found to be positive and significant. This means that for every 1 per cent increase in labour use, the gross income increased by 0.28 per cent in Arabica and 0.18 per cent in Robusta, respectively. The application of fertilizers (kg/ha) was found to be positive and significant in both cases, which was more prominent in Arabica. For every 1 per cent increase in fertilizer application, the gross income per ha of Arabica increased by 0.39 per cent, whereas in Robusta it increased by 0.15 per cent. Likewise, the cost of irrigation among both varieties of coffee was found to be significant in both cases; that is, for every 1 per cent increase in the cost of irrigation, the gross income per ha of Arabica increased by 0.01 per cent and for Robusta, it increased by 0.21 per cent. The application of pesticides had a positive influence on the gross income of Arabica; however, it was found to be insignificant in Robusta. In both varieties, the application of weedicides (litres/ha) was found to be insignificant.

Farm gate sales were commonly practised by coffee growers. About 31 per cent of Arabica and 52 per cent of Robusta growers sold their produce to local agents at the farm gate itself. The farm gate prices in India were higher than those in other coffee-producing countries, with Indian Robusta and Arabica growers receiving 92 per cent and 79.52 per cent of ICO indicator prices, respectively. In the case of Arabica parchment, the maximum price fluctuations occurred in the first quarter (January to March), followed by the fourth quarter. These two quarters also coincided with the onset of the procurement season for Arabica, wherein about 81.66 per cent of Arabica growers sold their produce that were subject to maximum price fluctuations and price risks.

In the case of Robusta cherry, the highest prices were observed in the fourth quarter (October to December), whereas the lowest prices were observed during the first quarter (January to March), which also coincided with the highest sales of Robusta, wherein about 83 percent of Robusta growers sold their produce. The highest farm gate prices for Arabica cherry were observed in the fourth quarter (October to December), specifically in the month of December.

About 30 per cent of Arabica growers sold their produce at the coffee depots, which were mere coffee-collecting centres. However, about 23 per cent of Arabica and 18 per cent of Robusta growers sold their produce to the curing works. About 15 per cent of Robusta growers sold their produce to hullers, where the milling process of coffee was carried out. Meanwhile, about 12 per cent of Arabica and Robusta growers sold certified and non-certified produce to companies. About 3 per cent of Arabica and 2 per cent of Robusta growers directly exported their produce, while only 2 per cent of Arabica growers were involved in online sales of coffee.

About 23.33 per cent of Arabica and 25 per cent of Robusta growers were certified. The Utz certification was commonly observed among the growers, followed by Rainforest Alliance (RFA) and Nespresso certification among Arabica. The Common Codes for the Coffee Community (4C) and organic certification were observed among the Robusta growers. However, only about 50 per cent of certified Arabica and 46.67 per cent of Robusta growers actually sold their produce as certified, while the rest was sold as non-certified coffee in the conventional market itself.

The price spread of Robusta cherry in market chain 1 (producers–agents–curers–roasters) brought out that the gross price received by the producer was ₹58.20 per kg. The expenses incurred by the producer were ₹1.40 per kg. The net margins incurred by agents, curers and roasters in channel 1 were ₹4.20, ₹5.27 and ₹125 per kg, respectively. The price spread was ₹241.80 per kg. The marketing costs and margins in channel 1 were ₹108.57 and ₹191.43, respectively.

Channel 2 (producers–hullers–curers and roasters-cum-retailers) showed that the gross price received by the producer for Robusta cherry was ₹59 per kg, which accounted for 19.66 per cent of roasters-cum-retailers' sales price. After deducting the expenses of ₹1.40 per kg, the producers received a net price of ₹57.60 per kg. The net margins incurred by hullers, curers and domestic roasters were ₹1.84, ₹2.50 and ₹131.53 per kg, respectively. The price spread was ₹241 per kg. The marketing costs and margins were ₹106 and ₹193.47 per kg, respectively.

Channel 3 for Robusta cherry comprised producers, curers and roasters-cum-retailers. The gross price received by the producer was relatively higher in this channel. After deducting the market expenses, the producer received a net price of ₹60.20 per kg, which was 20.06 per cent of the sales price of roasters-cum-retailers. The margin of curers and domestic roasters-cum-retailers was ₹4.27 and ₹128.80 per kg. The marketing costs and margins were ₹106.73 and ₹193.27, respectively. The price spread in this channel was ₹238 per kg.

Channels 4, 5 and 6 were related to Arabica parchment. Channel 4 comprised producers, agents, depots, curers and roasters. The price spread in this channel was ₹283.50 per kg. The producers' share in the sales price of roasters-cum-retailers was higher, viz. 37 per cent. The expenses incurred by the producers were ₹1.50 per kg. In this case, the growers received a net price of ₹165 per kg. The net margin of agents, coffee depots, curers and roasters-cum-retailers was found to be ₹1.50, ₹0.50, ₹8.72 and ₹72.56 per kg, respectively. The marketing costs and margins were higher, viz. ₹201.72 and ₹248.28, respectively.

Channel 5 comprised producers, coffee depots and roasters-cum-retailers. The gross price received by the producers was ₹168 per kg, while the costs incurred by the producer were ₹1.80, and the net price received by the producer was ₹166.20 per kg of Arabica parchment. The net margin of depots and roasters-cum-retailers was ₹0.70 and ₹92.89 per kg, respectively. The price spread was ₹282 per kg, while the selling price of roasters-cum-retailers was ₹450 per kg.

Channel 6 (producers–curers and roasters) showed that the growers received a net price of ₹167.10 after deducting ₹1.90 per kg. The gross price received by the curers was ₹183.32 per kg. The costs incurred by curers were ₹2.70 per kg, while the net margins received by them were ₹11.62 per kg. The domestic roasters-cum-retailers received a gross price of ₹450 per kg, while the net margin of roasters was ₹72.56 per kg. The price spread in this channel was found to be ₹281 per kg of Arabica parchment.

The marketing efficiency was measured in terms of modified Acharya's formula. This was found to be highest in channel 6 (0.38), followed by channels 5 and 4, where the marketing efficiency was found to be 0.37. The marketing efficiency for Robusta cherry showed the highest efficiency in channel 3, which was found to be 0.20, followed by channel 2 (0.196) and channel 1 (0.194). The higher ratio was associated with higher efficiency. Therefore, channel 4 (Producers–curers and roasters-cum-retailers) was found to be relatively efficient. The lowest marketing efficiency was observed in channel 1 in the case of Robusta.

The regression analysis on the factors affecting the net price received by the growers indicated that the net prices received by Robusta growers decreased with an increase in the area, which was found to be significant at a 5 per cent level of significance. The moisture content was found to be negative (–0.09 and –0.17 in Arabica and Robusta, respectively) and significant in both cases, which meant that the net prices of coffee reduced with an increase in the moisture content. The coefficient associated with out-turn was found to be significant and positive (0.75 and 1.97 in Arabica and Robusta, respectively). The costs associated with harvesting were found to be significant in Robusta but insignificant in Arabica. The total assets owned by the Robusta growers had a positive influence on the net prices of Robusta, viz. 1.89 which was significant at a 10 per cent level of significance.

The major problems faced by the producers in the maintenance of quality were species and genotypes (shift from Arabica to Robusta hampering quality), lower sustainability initiatives due to higher costs of certification, difficulties in obtaining inputs and a tedious recordkeeping process, coffee berry borer, lower out-turn and weight of berries, non-availability of resistant cultivars against leaf rust, berry borer disease and white stem borer, Rule 7 B of income tax of 1962 that discouraged value addition in coffee, type of harvesting methods due to non-availability of labourers, difficulties involved in selective harvesting and technical barriers (non-adoption of harvest mats, dry processing, mixing ripe and unripe berries, mud or cow-dung drying yards).

The major problems associated with traders and exporters were the limited repute of Indian coffee, contractual problems in fulfilling their contractual commitments, uncertainty in weather conditions during the shipment process, non-availability of the right package material, storage issues with coffee beetle, OTA contamination during shipment, difficulties in sourcing sustainable coffee, heavy dependence on agents and middlemen for procurement of coffee from the growers, low value addition among the exporters, pesticide residues in Indian coffee during exports and difficulties in the shipment of coffee.

The land tenure as an institution takes into account redeemed and unredeemed systems in Kodagu. The productivity of Arabica was higher in unredeemed land (813 kg/ha) than in redeemed land (798 kg/ha), while the shade tree density was also found to be higher in unredeemed Arabica estates (359 shade trees/ha) compared to redeemed Arabica estates (239 shade trees/ha). In contrast to Arabica, the productivity of Robusta was found to be higher in redeemed land (1715 kg/ha) compared to unredeemed land (1472 kg/ha). Furthermore, the fast-growing exotic silver oaks was higher in redeemed land (33 per cent and 62 per cent in redeemed Arabica and Robusta, respectively). Additionally, the shift from Arabica to Robusta variety of coffee was more pronounced in redeemed land.

The New Coffee Bill 2016 seeks to obtain more accurate estimate of the area under coffee and reliable inventories of the coffee-growing regions by withdrawing the Coffee Act of 1942. The New Bill had empowered the Coffee Board with powers of registration, formulation of strategies, maintenance of quality standards and issuing directives against pests and diseases. However, the New Coffee Bill 2016 appeared to be a mere replica of the Coffee Act of 1944. The growers and various associations had opposed the Bill, fearing the new bill would bring back the control system in the marketing of coffee. Additionally, the registration process and powers of Coffee Board officials to enter the premises of estates and take legal actions against the growers had created discontent among the growers. The proposed rationale of the New Coffee Bill 2016 still remains unclear in the coffee sector.

Prior to liberalization, the Coffee Board was involved in the promotion of domestic coffee markets, running demonstration vans, incentivizing people on coffee consumption, sales of coffee powder and liquid coffee at subsidized rates, advertisements on coffee, establishment of coffee houses, training on roasting, grinding and preparation of coffee and establishment of coffee houses, coffee bars, coffee vans and coffee sales depots. However, in 1996, the monopoly of the Coffee Board was broken and its role of Coffee Board was restricted to research, development and promotional activities.

Regarding CLR, the fundamental works on Coffee Leaf Rust disease were conducted in India. In order to control CLR, quarantine by-laws were regulated under the Destructive Insects and Pests Act of 1914, wherein the imported seeds were treated with systemic fungicides while complete prohibition of coffee beans or any other materials from Sri Lanka. The rust resistant Chandragiri variety was introduced in 2007 by CCRI, Balehonnur. The seeds of Chandragiri and other Arabica varieties (Sln.5A, Sln.5B, Sln.6 and Sln.9) were distributed in Karnataka and Tamil Nadu under the project "Increasing the resilience of coffee production to leaf rust and other diseases in India and four African countries".

With regard to Coffee Berry Borer, the four major institutes, viz. Coffee Board of India, Directorate of Plant Protection, Quarantine and Storage, UPASI and Directorate of Horticulture in Karnataka and Tamil Nadu, conducted a joint survey. Besides, innumerable extension activities were carried out by the Board from 1998–99 to 2001, which included visits to IPM plots, study tours, workshops, seminars, radio announcements and talks. While CBB was identified under section 4A of the DIP Act in 1992, legal restrictions were imposed on the movement of coffee from Kodagu, Nilgiri and Waynad. The two research projects emphasized educating the growers on CBB management while encouraged the growers to install berry borer traps.

In order to encourage the growers to take up eco-certification, the Coffee Board had provided subsidies to individual growers for up to 20 ha. In the case of organic certification, the growers/collectives were provided a subsidy covering 50 per cent of the certification costs with a ceiling of ₹50,000 per individual or group, while also providing technical support and capacity-building programmes to the growers, besides initiating the project 'Promotion of Organic Coffee Production in India'. The companies, as a part of Corporate Social Responsibility (CSR), support sustainable practices for eco-certification.

The Coffee Board explored new areas under coffee cultivation, viz. Araku Valley in Andhra Pradesh, Himachal Pradesh, Uttarakhand and West Bengal. The Araku Valley coffee was supported by the Coffee Board along with the assistance of the NGO Naandi Foundation and SKAL. In Himachal Pradesh, coffee cultivation was initiated in Bilaspur, Hamirpur, Mandi, Una, Kangra, Solan and Sirmaur districts, which resembled the foothills of the Mullayagiri range of Chickmagalur.

The IT application in coffee cultivation included the Kaapi Soil Health Management and Monitoring (Kshemam) app launched by the Indian Institute of Information Technology and Management-Kerala (IITM-K) to provide necessary information on soil management. Another initiative by the Coffee Board was the CAFE MOVEL project launched in 2013, which was based on Interactive Voice Response Technology (IVR). This helpline service was provided to the growers and included interactive learning on production practices, thereby alerting the growers on pests and diseases, weather conditions and prices. The producers were also subject to one-to-one interaction with experts.

The major growers' associations in the coffee sector were the United Planters' Association of Southern India (UPASI), Karnataka Planters' Association (KPA) and Codagu Planters' Association (CPA). The UPASI mainly covered scientific research and publications, information exchange, market intelligence, industrial public relations, training and estate supplies.

The KPA, being an apex decision-making body, made representation to concerned ministries on various issues of the industry. Likewise, the CPA aimed to promote and safeguard the plantation industry of Kodagu, represent policy decisions, analyze and publish statistics related to plantations and correspond with Planter's and Employer's Association, trade unions and the government.

The Coffee Futures Exchange of India (COFEI), National Multi-Commodity Exchange (NMCE), National Commodity and Derivative Exchange Limited (NCDEX) and Multi-Commodity Exchange (MCX) were involved in future trading of coffee. Meanwhile, ITC established e-choupals as an initiative to directly link coffee growers with companies for procurement of coffee. Online auctions were also introduced in coffee wherein the bids were announced and the produce was sold to the highest bidder.

The major associations involved in coffee processing and exports included the Indian Coffee Traders Association (ICTA), which conducted weekly coffee auctions every Thursday. Others included Speciality Coffee Association of India (SCAI), which seeks to expand sustainable speciality coffee trade through education, research and events to carve a niche for Indian coffee. This association also holds India Barista Championship and Coffee Mela. The All-India Coffee Curers Association (AICCA) represents the coffee curers, engaging in facilitating processing and policy dialogues. The Indian Coffee Roasters Association represents the interests of roasters, promoting differentiation strategy and high-quality segments of markets on the basis of blends and quality grades. It promotes speciality coffee roasters, high-tech roasting methods other than the traditional pan-frying method, value addition, development of local roasting capacity, training for roasters and soluble manufacturing capacities. However, the roasters and traders' associations lacked collective memberships and were fairly weak. The Coffee Exporters Association provided its members with commercially useful market information, market networks, trade fairs, access to special markets and technical advice.

Private institutions played a major role in the retailing of coffee. The major ones included Cafe Coffee Day (CCD), Barista, Qwiky and Starbucks. Meanwhile, the Coffee Board promotes domestic consumption of coffee through the launch of coffee stamps, Kaapi Shastra or training programmes in coffee and authorization of the registration of exporters; the major schemes for the promotion of exports included the Foreign Trade Policy and Market Development Assistance (MDA).

9.3 Suggestions

1. The Coffee Board, NGOs and other organizations need to sensitize coffee growers to replant and rejuvenate their old, senile and diseased trees in a bid to increase production. The Coffee Board should set a target of about 10 per cent of total area in the country to be replanted every year. Technical support along with financial support should be provided to the

growers. In order to maintain higher production and productivity, there is a need for timely replantation of old-aged coffee trees. Therefore, the organizations should systematically promote replanting through provision of subsidies and technical support.

2. The cultivation of Arabica coffee should be limited to suitable areas, viz. high elevation belt, so as to arrest the proliferation of WSB. In other words, Arabica cultivation should be encouraged only in estates situated above 1000 m.a.s.l.
3. Although there are several control measures for WSB, they turn out to be less effective and more costly. Several research works have been carried out in isolation in different countries, but no country has ever come up with an effective solution. A combined and coordinated inter-country research activity is required for sustainable management strategies.
4. To keep a check on White Stem Borer in Arabica, the development of pest-/disease-resistant varieties through a combination of antixenosis and antibiosis characteristics in coffee could prove effective. Other options include biological control of borers through *B. thuringiensis*.
5. Extensive campaigning and advertisement of certified coffee in domestic markets, besides creating a niche market for certified Kodagu coffee, could be beneficial. Here, the extension wing of the Coffee Board could play a vital role in encouraging growers to participate in the certification programmes.
6. The manufacturers can buy their coffee directly from growers and guarantee a minimum price for the growers.
7. If the small growers receive a lower price because they sell at farm gate, their welfare could be raised by offering institutional alternatives to farm gate sales, such as producers' co-operatives.

Annexure 1

Household questionnaire

I. Identification of household

Sr. no.	*Name of respondent*
Village	Taluk
District	Household size
Main product	Main market
Intercrops	e-mail address
Contact number	Education
Age	Years of experience
Land tenure system	

1. Estate profile

Occupation:	(a) Main:	(b) Subsidiary:
Elevation (ft or metres approx.) Slope		Annual rainfall (inches or mm) Soil type and pH

Slope codes (1 = Plain, 2 = Gentle slope, 3 = Medium slope, 4 = Steep slope, 5 = Very steep slope).

2. Family information

Sl. No	*Relation with head*	*Sex (1 = Male; 2 = Female)*	*Age (yr)*	*Marital status (1 = Married 2 = Unmarried)*	*Education status*	*Activity status*		*Net earnings from non-farming activities during 2015–16*
						Main occupation	*Subsidiary occupation*	
1								
2								
3								
4								
5								

(Code 1: 1 = Illiterate, 2 = Primary, 3 = Middle, 4 = High School, 5 = Graduate, 7 = Postgraduate and above).

(Code 2 : Activity status: 1 = Self-employed farming; 2 = Self-employed non-farming; 3 = Salary; 4 = Agricultural wages; 5 = Non-agricultural wages; 6 = Pensioner; 7 = Dependent; 8 = Household work; 9 = Student).

3 Particulars of land

Description	Distribution of area by source of irrigation (acres)						Total	Terms of leasing (Code 1)	In case of fixed rent total value of cash/kind paid/ received for leasing in/out
	Only canal	Canal+tubewell	Only electric tube well	Only diesel tube well	Tank and others	Rain fed area			

(Code 1: 1 = Share cropping, 2 = fixed rent in cash, 3 = fixed rent in kind; 4 = both; 5 = against labour; 6 = others).

4. Details of loan

Details of debt	*Amount borrowed (₹)*
Amount (₹)	
Source of borrowing (Code 1)	
Purpose of loan (Code 2)	
Year of borrowing	
Rate of interest (per cent per annum)	
Total debt outstanding loan (₹)	

(Code 1: 1 = Commercial banks; 2 = Co-operatives; 3 = RRBs; 4 = Commission agents; 5 = Coffee traders; 6 = others).

(Code 2 : 1 = Purchase tractor/tiller/other implements; 2 = Purchase of land; 3 = crop loan; 4 = Purchase of livestock; 5 = Daily consumption; 6 = Social ceremonies).

5. Particulars of assets

Assets	*Present Value (₹)*
Buildings	
Farm House	
Cattle House	
Store House	
(Details on days of storage)	
Drying Yard	
(Thickness of spreading)	
a) Mud/Cow dung yard	
(Reasons)	
b) Cement Yard	
c) Tiled	
Implements and Machinery	
Weeder	
Tractors and tillers	
Sprayer	
Sprinkler set	
Pulpers	

6. Variety, system and planted area of coffee by types

Coffee type	*Coffee variety (Code 1)*	*Current bearing area (acres)*	*Non-bearing (acres)*	*Year of establishment*	*Establishment cost*		
					1st year	*2nd year*	*3rd year*
Arabica							
Robusta							

(Code 1: S.795, 2 = Sln.5A, 3 = Sln.5B, 4 = Sln.6, 5 = Sln.9, 6 = Cauvery, 7 = Chandragiri).

Production and sales price by types

7. Intercrops in coffee

Crops	*Area (Acres)*	*Annual production (tonnes)*		*Annual income/sales revenue from crop (₹)*	
		2014–15	*2015–16*	*2014–15*	*2015–16*
Pepper					
Cardamom					
Areca					
Others (specify)					

8. Fertilizer costs and impacts on yield

Fertilizers	*Frequency*	*Months of application*	*Materials*		*Labour costs*	*Impact on yield*	*Any remarks*
			Qty (kg/acre)	Costs (₹/acre)			
Urea							
DAP							
MOP							
Liming							
Organic							
Others							

9. Pests and diseases and impacts on yield

9.1 Have you experienced any pests/diseases in your farm? Yes = 1, No = 2

If Yes in (12.1), specify pests/diseases (Circle answer)

- White Stem Borer = 1
- Coffee Leaf Rust = 2
- Black rot = 3
- Coffee Berry Borer = 4
- Shot Hole Borer = 5
- Others = 6

Pesticides	*Frequency of application*	*Timing of application (Code1)*	*Materials*		*Labour costs (₹/acre)*
			Qty (barrels/acre)	*Costs (₹/acre)*	
Bordeaux mix					
Cu fungicides					
Lindane					
Zn sulphate					
Neem					

9.2. Costs of application of pesticides

(Code1: Pre Blossom = 1; Pre-monsoon = 2; Post monsoon = 3)

9.3. Have you uprooted, replanted coffee plants recently (Past 3 years)? [Yes = 1, No = 2]

If Yes in (9.3), go to question 9.4. If no go to question 9.5

9.4 Uprooting, replanting varieties and corresponding cost

Uprooted varieties	*Replanted varieties*	*Source of purchase*	*Costs (₹/plant)*	*Perception on costs*

9.5 Coffee White Stem Borer (WSB) and impact on productivity

1. When did you realize WSB in your estate (Mention year)
2. How is the incidence of WSB from the time you noticed in your estate? (Circle answer) 1 = Increasing, 2 = Constant, 3 = Decreasing
3. Type of severity (Circle answer): Low = 1, Medium = 2, Severe = 3
4. Which are varieties are most affected by WSB (Circle answer) : S.795 = 1, Sln.5A = 2, Sln.5B = 3, Sln.6 = 4, Sln.9 = 5
5. What are the factors responsible for spread of WSB

- Improper shade management = 1
- Neglected neighbouring estates = 2
- Environmental changes = 3
- Near-by paddy fields = 4
- Others (specify) = 5

6. Under which type of shade regulation, WSB is more prevalent? (Circle answer)
   - Single tier = 1
   - Two tier = 2
   - Mixed shade = 3
   - Silver oak mixed shade = 4
7. How many plants you uproot per acre?
8. What is yield loss per acre (kg/acre)?
9. What are control measures you adapt to control WSB? (Circle)
   - Pheromone traps = 1
   - Chemical control = 2
   - Planting resistant varieties = 3
   - Others (Specify) = 4
10. What are your perceptions on effectiveness of control measures?
11. In your view, which season/month of the year is WSB most severe (circle answer)
    - January–March = 1
    - April–August = 2
    - September–December = 3
    - Any other (specify) = 4
12. Has White Stem Borer incidence prompted you to shift from Arabica to Robusta during last three years? Yes = 1, No = 2

12.1. Please provide your perceptions on the following areas

Attributes	*Do not agree*				*Strongly agree*
	1	*2*	*3*	*4*	*5*
Shifting towards Robusta					
12.1 I am shifting towards Robusta due to WSB					
12.2 I am shifting towards Robusta due to CLR					
12.3 I am shifting towards Robusta due to Labour shortage					
12.4 I am shifting towards Robusta due to higher Management costs					
Willing but unable to shift					
12.5. I unable to shift towards Robusta due to lack of irrigation					

Attributes	*Do not agree*				*Strongly agree*
	1	*2*	*3*	*4*	*5*
12.6. I am unable to shift towards Robusta due to unsuitability of land for Robusta					
No shift towards Robusta					
12.6 I do not want to shift to Robusta due to passion for Arabica					

9.6 Does Coffee Board provide any incentives for control of WSB?

10. Shade management and costs incurred

Coffee varieties	*Number of shade trees/ acre*	*% of shade cover*	*Shade tier (Code 1)*	*Shade management costs (₹/acre)*	*Impact of shade on yield*	*Impact of shade on pests and diseases*	*Growers perception on shade*
Arabica							
Robusta							

(Code 1 : One tier shade = 1; Two tier shade = 2 ; Three tier shade = 3)

11. Pruning practices and costs incurred

Coffee varieties	*Type of pruning*	*Frequency of pruning*	*Time of pruning*	*Material costs (₹)*	*Labour costs (₹/ acre)*	*Impact of pruning on yield*
Arabica						
Robusta						

(Code 1: Simple pruning = 1; Collar pruning = 2; Heavy pruning)

12. Irrigation application and costs

Coffee varieties	*Frequency of irrigation*	*Timing of irrigation*	*Labour costs(₹/ acre)*	*Impact of irrigation on yield*
Arabica				
Robusta				

13. Harvesting costs

Coffee varieties	*Type of harvest (Code 1)*	*Frequency of harvest*	*Start of harvest (month)*	*Labour costs/ acre*	*Material costs*
Arabica					
Robusta					

(Code 1: Fly picking = 1; Selective picking = 2; Stripping = 3)

14. After harvest, do you mix ripe berries with unripe berries?

15. Do you undertake gleaning activities? Yes = 1; No = 2

If Yes, go to question (16) and (17) . If no, go to question (21)

16. How many days after harvest, you undertake gleaning activities?

17. What are the costs incurred for gleaning?

18. Do you undertake weeding activities regularly? What are the costs incurred per acre of weeding?

19. Details of processing activities undertaken by the cultivator

Varieties processed	*Type of processors used*	*Final product (Quantities in kg)*	*Processing cost (₹)*	*Racking*			*Net profit (₹)*
				Method of raking	Frequency/ day	Costs (₹)	

1. Do you own a pulping unit? If yes, quantity processed?
2. If no, do you pulp it on rent? If yes cost incurred?

20. Marketing details

Type of coffee	*Place of market*	*Sold to (Code 2)*	*Qty sold (Bags)*	*Storage details*		*Packing costs (₹)*	*Transport costs (₹)*	*Other costs if any*
				Type	*Costs (₹)*			
Robusta cherry								
Robusta Parch								
Arabica cherry								
Arabica Parch								

(Code 2: Local Trader = 1; Companies = 2; Curing agencies = 3; Auction = 4; Exporters = 5; Brokers = 6; Retailers = 7; Direct export = 8; Any others = 9)

21. Quality parameters

Parameters	*Tests*	*Quality standards (per cent)*	*Price deductions (₹)*	*Premiums received (₹)*	*Growers perception*
1. Moisture content					
2. Out-turn per cent					
3. Berry Bore					
4. Blacks and bits					
5. Moulds					
6. Foreign matter					

22. Constraints in coffee production and marketing

Constraints	*Rank*		*Currently how are you trying to overcome the constraint*
	R1	R2	
Pests			
White Stem Borer			
Coffee Berry Borer			
Shot Hole Borer			
Diseases			
Coffee Leaf rust			
Other diseases			
Human–Animal Conflict			
Elephant menace			
Wild Borer			
Monkeys			
Other wild animals			
Production			
Labour shortage			
Inorganic fertilizers			
Farm Yard Manure			
Fungicides/pesticides			
Implements			
Market related			
Price related			
Transportation related			
Immediate payment			
Storage			
Price deductions			
Others(specify)			

R1 = Rank within individual groups; R2 = Overall rank among all constraints

23. General information about cultivation

1. Do face elephant menace in your estate? Yes = 1, No = 2
2. If Yes in (1), specify the impact on yield
   - Low = 1
   - Medium = 2
   - High = 3
3. What are the schemes of Coffee Board availed to you ? Kindly explain
4. Rather than farm-gate, do you sell coffee through some other agent?
5. Suggestions on production and marketing

List of annexure tables

Table 1.1a Trends in area, production and productivity of coffee in India

India						
Year	*Area (ha)*		*Production (MT)*		*Productivity (kg/ha)*	
	Arabica	*Robusta*	*Arabica*	*Robusta*	*Arabica*	*Robusta*
2008–09	156421 (44.63)	194079 (55.37)	79500 (30.31)	182800 (69.69)	508	942
2009–10	159828 (44.96)	195674 (55.04)	94600 (32.67)	195000 (67.33)	592	997
2010–11	163737 (45.42)	196748 (54.58)	94140 (31.17)	207860 (68.83)	575	1056
2011–12	169906 (46.08)	198781 (53.92)	101500 (32.32)	212500 (67.68)	597	1069
2012–13	176131 (46.81)	200174 (53.19)	98600 (30.99)	219600 (69.01)	560	1097
2013–14	181129 (47.50)	200175 (52.50)	102200 (33.56)	202300 (66.44)	564	1011
2014–15	185978 (48.16)	200217 (51.84)	98000 (29.97)	229000 (70.03)	527	1144
2015–16	192734 (48.53)	204413 (51.47)	103500 (29.74)	244500 (70.26)	537	1196
2016–17	195412 (47.67)	214512 (52.33)	95000 (30.45)	217000 (69.55)	486	1012
2017–18	198671 (48.10)	214349 (51.90)	95000 (30.06)	221000 (69.94)	478	1031
2018–19	202218 (48.52)	214523 (51.48)	95000 (29.73)	224500 (70.27)	470	1047
2019–20	203546 (48.68)	214621 (51.32)	87000 (29.19)	211000 (70.81)	427	983
2020–21	207060 (48.96)	215864 (51.04)	99000 (29.64)	235000 (70.36)	478	1089
Exp. GR	(2.49)	(1.01)	(0.37)	(1.38)	−(2.07)	(0.37)

Source: Indian Coffee Board.

Note: Figures in parentheses indicate percentage to total area and production.

Table 1.2a Trends in area, production and productivity of coffee in Karnataka

Karnataka						
Year	*Area (ha)*		*Production (MT)*		*Productivity (kg/ha)*	
	Arabica	*Robusta*	*Arabica*	*Robusta*	*Arabica*	*Robusta*
2008–09	110298 (48.61)	116602 (51.39)	81270 (37.95)	132900 (62.05)	749	1239
2009–10	98051 (47.88)	106752 (52.12)	73400 (35.68)	132300 (64.32)	749	1239
2010–11	109173 (47.94)	118577 (52.06)	79925 (36.39)	139700 (63.61)	732	1178
2011–12	109128 (47.52)	120530 (52.48)	79825 (36.12)	141175 (63.88)	731	1171
2012–13	109003 (47.32)	121330 (52.68)	77425 (33.63)	152800 (66.37)	772	1351
2013–14	109003 (47.32)	121330 (52.68)	78440 (37.16)	132660 (62.84)	782	1173
2014–15	109032 (47.32)	121402 (52.68)	74755 (32.05)	158475 (67.95)	745	1402
2015–16	111225 (47.24)	124213 (52.76)	78650 (31.27)	172870 (68.73)	775	1480
2016–17	108845 (44.47)	135940 (55.53)	71600 (31.64)	154700 (68.36)	707	1195
2017–18	108795 (44.45)	135990 (55.55)	69025 (31.05)	153275 (68.95)	692	1211
2018–19	108816 (44.36)	136472 (55.64)	67950 (30.95)	151600 (69.05)	682	1198
2019–20	108905 (44.33)	136777 (55.67)	60870 (29.92)	142575 (70.08)	609	1123
2020–21	99004 (43.59)	128098 (56.41)	73000 (31.00)	162500 (69.00)	737	1269
Exp. GR	−(0.07)	(1.65)	−(1.54)	(1.35)	−(0.96)	−(0.11)

Source: Indian Coffee Board.

Note: Figures in parentheses indicate percentage to total area and production.

Table 1.3a Trends in area, production and productivity of coffee in Kerala

Kerala						
Year	*Area (ha)*		*Production (MT)*		*Productivity (kg/ha)*	
	Arabica	*Robusta*	*Arabica*	*Robusta*	*Arabica*	*Robusta*
2008–09	3742 (4.42)	80954 (95.58)	1425 (2.49)	55775 (97.51)	381	719
2009–10	3602 (4.28)	80493 (95.72)	1375 (2.32)	57875 (97.68)	382	719
2010–11	3761 (4.43)	81170 (95.57)	1600 (2.43)	64175 (97.57)	425	791
2011–12	3865 (4.55)	81083 (95.45)	1900 (2.79)	66200 (97.21)	492	816
2012–13	4175 (4.89)	81184 (95.11)	2075 (3.23)	62125 (96.77)	537	771
2013–14	4175 (4.89)	81184 (95.11)	2000 (3.00)	64675 (97.00)	517	803
2014–15	4203 (4.92)	81253 (95.08)	2055 (3.04)	65645 (96.96)	532	815
2015–16	4217 (4.93)	81284 (95.07)	2200 (3.18)	67030 (96.82)	557	830
2016–17	4228 (4.92)	81642 (95.08)	2140 (3.38)	61150 (96.62)	541	754
2017–18	4231 (4.93)	81649 (95.07)	2160 (3.29)	63575 (96.71)	546	785
2018–19	4231 (4.93)	81649 (95.07)	2210 (3.14)	68225 (96.86)	559	842
2019–20	4231 (4.93)	81649 (95.07)	1800 (2.73)	64125 (97.27)	455	791
2020–21	3955 (4.65)	81021 (95.35)	1820 (2.63)	67280 (97.37)	460	830
Exp. GR	(1.06)	(0.07)	(2.59)	(0.97)	(1.94)	(0.81)

Source: Indian Coffee Board.

Note: Figures in parentheses indicate percentage to total area and production.

Table 1.4a Trends in area, production and productivity of coffee in Tamil Nadu

Tamil Nadu						
Year	*Area (ha)*		*Production (MT)*		*Productivity (kg/ha)*	
	Arabica	*Robusta*	*Arabica*	*Robusta*	*Arabica*	*Robusta*
2008–09	25708 (82.02)	5636 (17.98)	12425 (74.74)	4200 (25.26)	483	745
2009–10	24588 (81.58)	5553 (18.42)	14650 (75.71)	4700 (24.29)	596	846
2010–11	25739 (82.12)	5605 (17.88)	12000 (72.73)	4500 (27.27)	466	803
2011–12	25708 (82.02)	5636 (17.98)	13385 (72.94)	4965 (27.06)	521	881
2012–13	24461 (81.55)	5535 (18.45)	12800 (73.69)	4570 (26.31)	523	826
2013–14	25939 (82.23)	5605 (17.77)	13950 (74.30)	4825 (25.70)	538	861
2014–15	26058 (82.28)	5613 (17.72)	13150 (73.57)	4725 (26.43)	538	854
2015–16	29062 (83.20)	5870 (16.80)	12810 (74.07)	4485 (25.93)	478	783
2016–17	29513 (82.89)	6094 (17.11)	12060 (72.83)	4500 (27.17)	427	772
2017–18	29513 (82.89)	6094 (17.11)	13400 (76.83)	4040 (23.17)	482	695
2018–19	29324 (82.39)	6268 (17.61)	13240 (74.53)	4525 (25.47)	481	760
2019–20	29338 (82.29)	6314 (17.71)	13225 (76.01)	4175 (23.99)	480	695
2020–21	27900 (82.29)	6004 (17.71)	12800 (71.51)	5100 (28.49)	459	849
Exp. GR	(1.55)	(1.07)	−(0.08)	−(0.07)	−(1.10)	−(0.74)

Source: Indian Coffee Board.

Table 1.5a Major states share in all India coffee area and production

State	*Share in all India*					
	Area (%)		*Production (%)*		*Total area %*	*Total production %*
	Arabica	*Robusta*	*Arabica*	*Robusta*		
Karnataka	59.0	60.9	78.0	68.8	59.9	71.6
Kerala	2.2	39.7	2.0	29.6	22.0	21.1
Tamil Nadu	14.8	2.8	13.7	2.1	8.5	5.7

Source: Indian Coffee Board.
(Average of 2008–09 to 2020–21)

Table 1.6a Trends in area, production and productivity of coffee in Kodagu

Kodagu						
Year	*Area (ha)*		*Production (MT)*		*Productivity (kg/ha)*	
	Arabica	*Robusta*	*Arabica*	*Robusta*	*Arabica*	*Robusta*
2008–09	28828 (27.90)	74497 (72.10)	19435 (18.26)	86975 (81.74)	674	1167
2009–10	28803 (27.81)	74777 (72.19)	21550 (19.38)	89650 (80.62)	748	1199
2010–11	28303 (27.32)	75277 (72.68)	22050 (19.55)	90750 (80.45)	779	1206
2011–12	28053 (26.77)	76727 (73.23)	21800 (18.95)	93225 (81.05)	777	1215
2012–13	27963 (26.66)	76927 (73.34)	21300 (17.75)	98700 (82.25)	762	1283
2013–14	27963 (26.66)	76927 (73.34)	21040 (18.81)	90820 (81.19)	752	1181
2014–15	27969 (26.66)	76953 (73.34)	18030 (14.76)	104090 (85.24)	676	1406
2015–16	28104 (26.38)	78423 (73.62)	19150 (13.95)	118150 (86.06)	715	1575
2016–17	26881 (26.33)	75199 (73.67)	17400 (14.02)	106750 (85.98)	634	1367
2017–18	28140 (26.28)	78949 (73.72)	19550 (16.77)	97000 (83.23)	727	1290
2018–19	28182 (26.27)	79114 (73.73)	16900 (15.26)	93830 (84.74)	630	1248
2019–20	28422 (26.42)	79144 (73.58)	17100 (15.96)	90075 (84.04)	638	1197
2020–21	26351 (25.90)	75391 (74.10)	20600 (16.49)	104350 (83.51)	782	1384
Exp. GR	-(0.38)	(0.34)	-(1.47)	(0.99)	-(0.75)	(0.96)
Kodagu's share in Karnataka	**Area (%)**		**Production (%)**		**Total area %**	**Total prod. %**
	Arabica	**Robusta**	**Arabica**	**Robusta**		
	26.0	61.7	26.5	65.6	45.1	52.5

Source: Indian Coffee Board.

Note: Figures in parentheses indicate percentage to total area and production.

Table 2.1a Distribution of sample households by size of land holdings

Size of holdings	*Arabica (N = 60)*	*Robusta (N = 60)*
Small (<2 ha)	8 (13.33)	8 (13.33)
Semi-medium (2–10 ha)	31 (51.67)	34 (56.67)
Medium (10–25 ha)	13 (21.67)	12 (20.00)
Large (>25 ha)	8 (13.33)	6 (10.00)
Total	**60** **(100.00)**	**60** **(100.00)**

Source: Field survey.

Table 2.2a The general description of the study area

Sl. no	*Particulars*	*Madikeri*	*Somwarpet*	*Virajpet*	*Overall*
1	Geographical area (ha)	145045	99999	165731	410775
2	Forest area (ha)	47514	20849	66234	134597
5	Elevation (metres)	1170	1130	909	1070
6	Rainfall (mm in 2015)	3492	1933	1683	2431
7	Villages (number)	67	135	94	296
8	Population (Lakh in 2011 Census)	1.46	2.06	2.01	5.53
	Male ('000)	0.72	1.02	1.00	2.74
	Female ('000)	0.74	1.04	1.02	2.80
9	Sex ratio 2011	1009	1027	1019	1019
10	Literacy rate (%)	88.2	81.5	79.6	82.66
	Male (%)	91.7	87.4	83.6	87.19
	Female (%)	84.7	75.8	75.7	78.14

Source: Kodagu District at a glance, 2015–16.

Table 3.1a Socio-economic profile of Arabica and Robusta growers

Arabica						*Robusta*				
Characteristics	*Small (N = 8)*	*Semi-medium (N = 31)*	*Medium (N = 13)*	*Large (N = 8)*	*Total (N = 60)*	*Small (N = 8)*	*Semi-medium (N = 34)*	*Medium (N = 12)*	*Large (N = 6)*	*Total (N = 60)*
Landholdings (ha)	1.57	5.19	14.83	90.4	18.15	1.10	5.60	18.40	45.66	11.56
Household size (numbers)	6.5	5.58	4.07	3.62	5.12	6.50	4.64	4.00	4.00	4.70
Average earners (numbers)	2.87	2.61	2.00	1.62	2.38	2.87	2.38	2.08	2.16	2.36
Per cent of earners (%)	44.23	46.80	49.06	44.83	46.58	44.23	51.27	52.08	54.17	50.35
Male (%)	60.87	63.00	69.23	84.62	65.73	60.87	62.96	80.00	76.92	66.90
Female (%)	39.13	37.00	30.77	15.38	34.27	39.13	37.04	20.00	23.08	33.10
Age group of the household members (%)										
<16 yrs	23.08	26.00	11.32	20.69	22.47	26.93	20.89	16.67	12.50	20.57
16–60 yrs	48.07	53.20	67.92	65.52	56.03	46.15	60.12	66.67	66.67	59.22
>60 yrs	28.85	20.80	20.76	13.79	21.50	26.92	18.99	16.66	20.83	20.21
Education status of growers (%)										
1 = Illiterate	0	0	0	0	0	0	0	0	0	0
2 = LP	12.50	3.23	0	0	3.33	0	0	0	0	0
3 = U P	37.50	0	0	0	5.00	12.50	0	0	0	1.67
4 = High School	37.5	9.68	0	12.50	11.67	25.00	5.88	16.67	16.67	11.67
5 = PUC	12.50	12.90	23.08	12.50	15.00	25.00	29.41	16.67	0	23.33
6 = Graduate	0	67.70	76.92	75.00	61.67	37.50	50.00	58.33	66.67	51.67
7 = PG	0	6.45	0	0	3.33	0	14.71	8.33	16.67	11.67

(Continued)

Table 3.1a Continued

Arabica						*Robusta*				
Characteristics	*Small (N = 8)*	*Semi-medium (N = 31)*	*Medium (N = 13)*	*Large (N = 8)*	*Total (N = 60)*	*Small (N = 8)*	*Semi-medium (N = 34)*	*Medium (N = 12)*	*Large (N = 6)*	*Total (N = 60)*
Caste of households (%)										
OBC	75.00	64.50	15.39	0	46.67	62.50	61.76	8.33	0	45.00
General	25.00	32.30	76.92	87.50	48.33	37.50	29.41	75.00	100	46.67
Minorities	0	3.23	7.69	12.50	5.00	0	8.83	16.67	0	8.33
Experience (years)	26.25	22.00	32.15	19.75	24.48	36.25	30.50	32.00	36.16	32.13

Source: Field survey.

Table 3.2a Estate profile of surveyed Arabica and Robusta growers

Estate profile	Arabica (% of estates)					Robusta (% of estates)				
	Small (N = 8)	*Semi-medium (N = 31)*	*Medium (N = 13)*	*Large (N = 8)*	*Total (N = 60)*	*Small (N = 8)*	*Semi-medium (N = 34)*	*Medium (N = 12)*	*Large (N = 6)*	*Total (N = 60)*
Elevation										
<1000 m	87.50	80.64	84.62	87.50	83.33	100	91.18	100	83.34	93.33
>1000 m	12.50	19.36	15.38	12.50	16.67	0	8.82	0	16.66	6.67
Annual Rainfall during reference year 2015–16										
1016–1778 mm	38	55	62	63	57	25	35	50	17	35
1778–2794 mm	63	45	38	38	43	25	24	0	50	22
27945080 mm–	0	0	0	0	0	50	41	50	33	43
Age of the plant (yr)										
15–20	37.50	21.05	0	16.66	18.33	25.00	27.02	25.00	16.66	25.00
21–25	37.50	21.05	44.44	0	26.67	0	10.54	0	16.66	6.66
26–30	12.50	23.68	22.23	66.67	26.67	50.00	24.32	25.00	0	26.66
31–35	0	23.68	11.11	0	13.33	0	8.40	12.50	16.66	10.00
36–40	12.50	7.89	11.11	16.67	11.67	25.00	27.02	0	16.66	20.02
>40	0	2.65	11.11	0	3.33	0	2.70	37.50	33.36	11.66
Land tenure										
Redeemed	75.00	70.96	69.23	50.00	68.33	50.00	50.00	41.66	50.00	48.33
Unredeemed	25.00	29.04	30.77	50.00	31.67	50.00	50.00	58.34	50.00	51.67

Source: Field survey.

Table 3.3a Intercrops in Arabica and Robusta coffee estates in Kodagu (2015–16)

Intercrops	*Area, productivity and gross revenue*	*Arabica*					*Robusta*				
		Small	*Semi-medium*	*Medium*	*Large*	*Total*	*Small*	*Semi-medium*	*Medium*	*Large*	*Total*
Pepper	Area (ha)	1.1	3.29	11.26	25.25	7.65	1.46	3.58	12.52	30	7.73
	Productivity (kg/ha)	175	197	149	165	180	200	241	197	200	223
	Gross revenue (₹/ha)	105718	133016	93317	95468	115768	121562	147481	121666	124791	136593
Areca	Area (ha)	0.15	0.61	0.74	1.5	0.69	0.18	1.02	0.8	1.58	0.92
	Productivity (kg/ha)	1123	1816	3334	4492	2409	2246	2478	3007	7113	3016
	Gross revenue (₹/ha)	22462	36326	66696	89850	48194	44925	49576	60149	142262	60339
Cardamom	Area (ha)	0	0.38	0.42	0.56	0.36	0	0.1	0.12	0.17	0.11
	Productivity (kg/ha)	0	64.47	75.49	138.43	73.68	0	45.36	57.17	116.67	58.02
	Gross revenue (₹/ha)	0	25787	30195	55371	29475	0	18143	22866	46666	23208
Coorg Mandarins	Area (ha)	0	0.98	1.15	2.57	1.1	0	1.07	2.81	3.99	1.57
	Productivity (kg/ha)	0	573	552	534	487	0	943	940	932	815

(Continued)

Table 3.3a Continued

Intercrops	*Area, productivity and gross revenue*	*Arabica*					*Robusta*				
		Small	*Semi-medium*	*Medium*	*Large*	*Total*	*Small*	*Semi-medium*	*Medium*	*Large*	*Total*
	Gross revenue (₹/ha)	0	8599	8284	8011	7306	0	14149	14107	13985	12238
Sapotas	Area (ha)	0	0.49	0.56	1.16	0.53	0.3	1.33	2	2.83	1.47
	Productivity (kg/ha)	0	464	1061	1675	693	1296	1259	1166	1216	1241
	Gross revenue (₹/ha)	0	6960	18535	25125	10396	19453	18893	17500	18250	18625

Source: Field survey.

Table 3.4a Total assets of Arabica and Robusta growers

Arabica (₹Lakhs)						*Robusta (₹Lakhs)*				
Farm size	*Small*	*Semi-medium*	*Medium*	*Large*	*Overall*	*Small*	*Semi-medium*	*Medium*	*Large*	*Overall*
Buildings	6.31	17.03	26.15	49.5	24.75	11.25	24.02	38.08	47.5	30.21
Labour line[a]	1.00	4.08	5.23	7.65	4.49	0.23	5.82	6.00	11.83	5.97
Cattle house	0.18	0.08	0.3	0.25	0.20	0.2	0.3	0.66	0.16	0.33
Store house	1.75	3.51	3.69	6	3.74	1.06	4.88	8.5	15.5	7.49
Mud/cow dung yard	0.37	0.03	0	0	0.10	0.75	0.17	0	0	0.23
Cement/tiled	2.18	6.16	7	32	11.84	0.7	6.91	10.91	26.66	11.30
Weeder	0.01	0.02	0.17	0.18	0.10	0.04	0.05	0.12	0.15	0.09
Tractors & tillers	0.81	0.91	1.92	11.75	3.85	0.62	1.12	5.6	14.08	5.36
Sprayer	0.03	0.03	0.041	0.08	0.05	0.02	0.02	0.05	0.06	0.04
Sprinkler set	0.12	0.69	1.23	5.62	1.92	0.23	2.16	2.41	5.16	2.49
Pulper	1.87	6.9	7	14.37	7.54	0	0.29	1.58	11.66	3.38
Mechanized dryer	0	0	0	0	0.00	0	0	0	16.66	4.17
Total Assets	14.63	39.61	52.63	127.4	58.57	15.1	45.74	68.28	132.76	65.47

Source: Field survey.

[a] Labour lines within the estates were provided for plantation workers for their stay.

Table 3.5a Credit details of Arabica and Robusta growers

	ARABICA					ROBUSTA				
Credit details	*Small (N=8)*	*Semi-medium (N=31)*	*Medium (N=13)*	*Large (N=8)*	*Total (N=60)*	*Small (N=8)*	*Semi-medium (N=34)*	*Medium (N=12)*	*Large (N=6)*	*Total (N=60)*
Growers with bank loans (%)	100	93.55	92.31	50.00	88.33	62.50	79.41	66.67	66.67	73.33
Amount borrowed (₹ in Lakhs)	2.16	3.35	4.88	33.75	7.58	0.60	4.22	6.87	23.33	6.18
Source of Borrowing (%)										
Commercial Banks	0	6.45	23.07	50.00	15.00	0	26.47	41.67	66.66	30
Co-Operatives	100	87.1	69.24	0	73.33	62.50	52.95	25.00	0	43.33
No debt	0	6.45	7.69	50.00	11.67	37.50	20.58	33.33	33.34	26.67
Purpose of loan (%)										
Crop Loan	100	97.06	100	100	98.33	100	100	83.34	100	96.67
Purchase of implements	0	0	0	0	0	0	0	16.66	0	3.33
Purchase of livestock	0	2.94	0	0	1.67	0	0	0	0	0

Source: Field survey.

Table 4.1a Descriptive statistics on productivity (kg/ha) of coffee

Statistical variables	*Arabica* (N = *60)*	*Robusta* (N = *60)*	*Overall* (N = *120)*
Mean	803	1590	1197
Minimum	600	1250	925
Maximum	1200	1800	1500
Standard deviation	143	156	406

Source: Field survey.

Table 4.2a Productivity (kg/ha) of coffee over years

Year	*All India*			*Karnataka*			*Kodagu*		
	Arabica (kg/ha)	*Robusta (kg/ha)*	*Ratio of R:A*[a]	*Arabica (kg/ha)*	*Robusta (kg/ha)*	*Ratio of R:A*	*Arabica (kg/ha)*	*Robusta (kg/ha)*	*Ratio of R:A*
2008–09	508	942	1.85	749	1,239	1.65	674	1167	1.73
2009–10	592	997	1.68	749	1,239	1.65	748	1199	1.60
2010–11	575	1056	1.84	732	1178	1.61	779	1206	1.55
2011–12	597	1069	1.79	731	1171	1.60	777	1215	1.56
2012–13	560	1097	1.96	772	1351	1.75	762	1283	1.68
2013–14	564	1011	1.79	782	1173	1.50	752	1181	1.57
2014–15	527	1144	2.17	745	1402	1.88	676	1406	2.08
2015–16	537	1196	2.23	775	1480	1.91	715	1575	2.20
2016–17	486	1012	2.08	707	1195	1.69	634	1367	2.16
2017–18	478	1031	2.16	692	1211	1.75	727	1290	1.77
2018–19	470	1047	2.23	682	1198	1.76	630	1248	1.98
2019–20	427	983	2.30	609	1123	1.84	638	1197	1.88
2020–21	478	1089	2.28	737	1269	1.72	782	1384	1.77

Source: Indian Coffee Board (2017).
[a] R:A – Robusta:Arabica

Table 4.3a Descriptive statistics on age of coffee trees (years) among the surveyed estates

Age of coffee trees (years)			
Statistical variables	*Arabica (n=60)*	*Robusta (n=60)*	*Total (N=120)*
Mean	28.02	30.58	29.30
Minimum	15.00	10.00	12.50
Maximum	46.00	60.00	53.00
Standard Deviation	7.87	10.28	9.07

Source: Field survey.

Table 4.4 Productivity (kg/ha) of coffee under different age group of coffee trees

Age of Arabica trees (years)	*Arabica (kg/ha)*	*Age of Robusta trees (years)*	*Overall (kg/ha)*
10-20 years	765	10-15 years	1350
21-25 years	946	16-25 years	1642
26-30 years	894	26-35 years	1718
31-35 years	733	36-40 years	1500
36-40 years	666	>40 years	1385
>40 years	600	-	

Source: Field survey.

Table 4.5a Coffee cultivation under different elevations in India

Region	*State*	*Elevation m.a.s.l.*	*Area under coffee (ha)*
Anamalais	Tamil Nadu	1000–1400	2500
Araku Valley	Andhra Pradesh	900–1100	20000
Bababudangiris	Karnataka	1000–1500	15000
Biligiris	Karnataka/TN	1500–2000	800
Brahmaputra	NE States	800–1200	5000
Chickmagalur	Karnataka	700–1200	60000
Coorg	Karnataka	750–1100	82000
Manjarabad/Sakleshpur	Karnataka	900–1100	41100
Nilgiris	Tamil Nadu	900–1400	7600
Pulneys	Tamil Nadu	600–2000	14000
Shevaroys	Tamil Nadu	900–1500	5000
Travancore	Kerala	400–1600	13000
Waynad	Kerala	600–900	67000

Source: Indian Coffee Board, 2017.

Table 4.6a Descriptive statistics on elevation of sample estates

Elevation of sample coffee estates (m.a.s.l.)			
Statistical variables	*Arabica (n = 60)*	*Robusta (n = 60)*	*Total (n = 120)*
Mean	947	924	936
Minimum	853	704	778
Maximum	1026	1072	1049
Standard deviation	53	119	71

Source: Field survey.

Table 4.7a Descriptive statistics on rainfall (mm) of sample estates during 2015–16

Rainfall (mm) in 2015–16[a]			
Statistical variables	*Arabica (N = 60)*	*Robusta (N = 60)*	*Total (N = 120)*
Mean	1866	2578	2222
Minimum	1397	1058	1229
Maximum	2413	4656	3534
Standard deviation	335	1205	770

Source: Field survey.

[a] Rainfall details are scientific as the growers accurately maintain the rainfall records.

Table 4.8a Productivity (kg/ha) of coffee under different rainfall ranges

Arabica			*Robusta*
Annual rainfall (mm)	*Productivity (kg/ha)*	*Annual rainfall (mm)*	*Productivity (kg/ha)*
1397–1524	769	<1500	1572
1525–1778	861	1501–2000	1612
1779–1905	825	2001–3000	1619
1906–2159	761	3001–4000	1617
2160–2413	766	>4000	1540

Source: Field survey.

Table 4.9a Optimum rainfall condition (mm) for Arabica as documented in earlier literature

Literature	*Optimum rainfall (mm) for Arabica*
Alegre, 1959	1200–1800 mm
Purseglove, 1968	1524–2286 mm
Hayward, 1987	1900 mm
Teketay, 1999	762–2540 mm
Taye et al., 2006	1500–1800 mm
Hermann et al., 2010	1400–2400 mm
Chemura, 2014	1800 mm

Source: Compiled from the literature.

Table 4.10a Ideal rainfall (mm) conditions for Arabica and Robusta varieties of coffee

Blossom showers		
	Arabica	*Robusta*
Months of ideal rainfall	March–April	February–March
Quantity of ideal showers	25–40 mm	25–40 mm
Backing showers		
Months of ideal rainfall	April–May	March–April
Quantity of ideal showers	50–75 mm	50–75 mm

Source: Indian Coffee Board, 2017.

Table 4.11a Critical rains (mm) vs productivity of coffee (kg/ha)

Arabica		*Robusta*	
Critical rains (mm)	*Productivity (kg/ha)*	*Critical rains*	*Productivity (kg/ha)*
150–200	813	<100	1531
200–250	800	101–125	1554
251–300	755	126–150	1657
>300	820	>150	1728

Source: Field survey.

Table 4.12a Descriptive statistics on shade tree density in sample estates

Number of shade trees per ha			
Statistical variables	*Arabica (n = 60)*	*Robusta (n = 60)*	*Total (N = 120)*
Mean	237	147	192
Minimum	100	100	100
Maximum	450	300	375
Standard deviation	89.89	50.30	70.10

Source: Field survey.

Table 4.13a Productivity (kg/ha) of coffee according to number of shade trees/ha

Arabica		*Robusta*	
Shade trees/ha	*Productivity (kg/ha)*	*Shade trees/ha*	*Productivity (kg/ha)*
<200	792	<150	1615
200–299	806	150–199	1630
300–399	810	200–250	1511
>400	783	>250	1275

Source: Field survey.

Table 4.14a Descriptive statistics on area under coffee (ha) of sample estates

Area under coffee (ha)			
Statistical variables	*Arabica (n = 60)*	*Robusta (n = 60)*	*Total (N = 120)*
Mean	18.16	11.57	14.86
Minimum	1.20	0.60	0.90
Maximum	158.40	72.00	115.20
Standard deviation	33.70	13.93	23.82

Source: Field survey.

Table 4.15a Productivity of coffee under different size groups

Area (ha)	*Productivity (kg/ha)*	*Area (ha)*	*Productivity (kg/ha)*
<2	775	<2	1575
2–10	808	2–10	1588
11–25	800	10–20	1606
26–50	833	21–50	1616
50–100	800	>50	1550
100–150	812		

Source: Field survey.

Table 4.16a Factors affecting productivity of Arabica: An overall regression analysis (Y = kg/ha)

Variable	*Coefficient*	*'t' value*	*Level of significance*
Constant	−261	−1.43	0.15
Age of coffee trees (years)	29.07	3.05***	0.003
Age square of coffee trees	−0.59	−3.46***	0.001
Elevation (m)	0.36	2.53***	0.01
Critical rains (mm)	−0.19	−0.93	0.35
Average annual rainfall (mm)	0.04	1.36	0.17
Area under coffee (ha)	0.65	2.05**	0.04
Shade trees per ha	0.24	2.34***	0.01
WSB incidence (uprooted plants per ha)	−7.66	−2.42***	0.01
Pruning (₹/ha)	0.04	6.54***	3.09*E*−08
R^2	0.80		
F-value	23.08***		
No. of observations	60		

Source: Field survey.
Note: *** and ** significant at 0.01 and 0.05 level, respectively.

Table 4.17a Inflection points in Arabica and Robusta

Inflection on age of coffee trees[a]		
Variable	*Formula*	*Inflection point*
Age of Arabica	$\beta_1/2\beta_2$	24.63 years
Age of Robusta	$\beta_1/2\beta_2$	30.81 years

Source: Authors' calculation.
[a] $\delta_y / \delta_{age} = \beta_1 - \beta_2*2(age)$ or
Age = $\beta_1/2\beta_2$
= 29.07/(2*0.59)= 24.63 yearsInflection point was estimated only when the coefficient associated with variables and it's square terms were significant.

Table 5.1a Details of growers shifted from Arabica to Robusta (N = 60)

Variable	*Shifting towards Robusta*	*Willing to shift, but unable to shift*	*Not shifting towards Robusta*
Small (<10 ha)	38.33	18.33	5.00
Large (>10 ha)	30.00	3.34	5.00
Total	68.33	21.67	10.00

Source: Field survey.[a]

[a] Note: All Figures refer to share (%) of farmers to total sample size. For instance 38.33 means that 38.33 per cent of 60 Arabica growers in Kodagu district. *N* refers to total sample size.

Table 5.2a Estate elevation (m.a.s.l.) and rainfall (mm) vs WSB incidence

Number of estates[a]						
WSB incidence	*LE + LR*[1]	*LE + MR*[2]	*LE + HR*[3]	*HE + LR*[4]	*HE + MR*[5]	*HE + HR*[6]
Low (<25 uproot plants/ha)	4 (7.00)	-	-	1 (2.00)	-	-
Medium (25–62 uproot plants/ha)	8 (13.00)	3 (5.00)	2 (3.00)	2 (3.00)	-	-
High (62–125 uproot plants/ha)	11 (18.00)	7 (12.00)	1 (2.00)	1 (2.00)	1 (2.00)	-
Very high (125–250 uproot plants/ha)	12 (20.00)	2 (3.00)	-	2 (3.00)	1 (2.00)	-
Critical (250–375 uproot plants/ha)	-	-	-	-	-	-

Number of estates[a]						
WSB incidence	*LE + LR[1]*	*LE + MR[2]*	*LE + HR[3]*	*HE + LR[4]*	*HE + MR[5]*	*HE + HR[6]*
Alarming (>375 uproot plants/ha)	1 (2.00)	1 (2.00)	-	-	-	-
Total estates	36 (60.00)	13 (22.00)	3 (5.00)	6 (10.00)	2 (3.00)	-

Source: Field survey.

[1] LE (Low Elevation) + LR (Low Rainfall) under <1000 m elevation & mean annual rainfall of 1397–1524 mm.

[2] LE (Low Elevation) +MR (Medium Rainfall) under <1000 m elevation & mean annual rainfall of 1524–2032 mm.

[3] LE (Low Elevation) + HR (High Rainfall) under <1000 m elevation & mean annual rainfall of 2032–2413 mm.

[4] HE (High Elevation) + LR (Low Rainfall) under >1000 m elevation & mean annual rainfall of 1397–1524 mm.

[5] HE (High Elevation) + MR (Medium Rainfall) under >1000 m & mean annual rainfall of 1524–2032 mm.

[6] HE (High Elevation) + HR (High Rainfall) under >1000 m & mean annual rainfall of 2032—2413 mm

Note: Figures in parentheses indicate percentage to total number of estates.

Table 5.3a Factors affecting uprooting of Arabica: Regression analysis (*Y* = uprooted trees/ha)

Variable	*Coefficient*	*'t' value*	*Level of significance*
Constant	80.124	–8.9***	0.00
Ln area under coffee (ha)	0.04	–0.7	0.50
Ln elevation (m.a.s.l.)	–8.365	–6.7***	0.00
Ln Actual rainfall (mm)	–0.395	–1.6	0.11
Ln shade trees (number per ha)	–0.409	–2.3**	0.03
Ln expenditure on pesticides (₹/ha)	–0.511	–2.2**	0.03
Dummy on age of plant (20–30 years)	–0.068	–0.4	0.66
Dummy age of plant (≥30 years)	–0.431	–2.2**	0.03
R^2	0.71		
F	21.13***		
Number of observations	60		

Source: Field survey.

*** and ** significant at 0.01 and 0.05 level, respectively.

Table 5.4a Cultural costs for White Stem Borer (WSB) control in Arabica (₹/ha)

Rate of incidence	*Uprooted (Arabica plants/ha)*	*Cultural Costs (₹/ha) WSB detection, uprooting and burning of stems*
Low	<25	670
Medium	25–62	875
High	62–125	1750
Very high	125–250	3625
Critical	250–375	4500
Alarming	>375	7500

Source: Field survey.

Table 5.5a Chemical costs for White Stem Borer (WSB) control in Arabica (₹/ha)

Chemicals	*Per cent of growers adopting chemicals*	*Material cost (₹/ha)*	*Labour costs (₹/ha)*	*Total costs (₹/ha)*
Chloropyrifos	56%	5393	2032	7425
Chloro + Lime	10.90%	5538	2500	8038
Lindane	23%	6045	2327	8372
Lindane + Contof	2.50%	6500	2500	9000
Lindane + Neem	2.50%	6250	2500	8750
Non-adoption	5.10%	0	0	0

Source: Field survey.

Table 6.1a Net returns per hectare from Arabica variety of coffee in 2015–16 (₹/ha)

Items	*Small*	*Semi-medium*	*Medium*	*Large*	*Overall*
1.Total Operational Cost (TOC)	**92232 (65)***	**100356 (62)***	**107964 (61)***	**112847 (59)***	**102559 (62)***
A. Labour Costs					
A1. Hired	10037	27597	47405	54175	33091
A2. Family	43072	29527	9582	5053	23748
Sub Total	53109 (58)**	57124 (57)**	56987 (53)**	59228 (52)**	56839 (55)**
B. Input Costs					
1.Manures and fertilizers	17694	18583	20180	20492	19065
2. Liming and dolomite	0	367	568	542	385
3. Irrigation	4201	5401	5742	6998	5528
4. Weeding	600	1070	1934	2749	1418
5. Pesticides	5695	5019	5649	5870	5359
Sub Total	28190 (31)**	30440 (30)**	34073 (31)**	36651 (32)**	31755 (31)**
C. Maintenance Costs					
1.. Transplanting and gap filling	1694	1325	1999	1378	1529
2. Fence keeping	130	180	471	168.00	235.00
Sub Total	1824 **(2)****	1505 **(1)****	2470 **(2)****	1546 **(1)****	1764 **(2)****
D. Processing Costs					
1. Processing costs	1755	1781	1983	1923	1841
2. Costs of materials	877	889	994	965	920
Sub Total	2632 (3)**	2670 (3)**	2977 (3)**	2888 (3)**	2761 (3)**
E. Marketing Costs					
1. Transportation charges	336	362	384	457	376
2. Storage charges	0	118	119	169	109
3. Labour charges	115	112	128	130	119
4. Value of bags	1154	1006	1076	1096	1026
Sub Total	1605 (2)**	1598 (2)**	1707 (2)**	1852 (2)**	1630 (2)**
F. Working capital of Arabica growers (₹)					
Total Working Capital (A + B + C + D + E - A2)	44288	63810	88632	97112	71001

(Continued)

Table 6.1a Continued

Items	*Small*	*Semi-medium*	*Medium*	*Large*	*Overall*
G. Interest on working capital at 11%	**4872**	**7019**	**9750**	**10682**	**7810**
II. Total Fixed Costs	**50049 (35)***	**62085 (38)***	**68646 (39)***	**78863 (41)***	**64138 (38)***
A. Rent on owned land	26396 (52.74)***	27658 (44.54)***	27251 (39.69)***	27514 (34.88)***	27382 (42.69)***
B. Land revenue, cesses and taxes	38 (0.07)***	38 (0.06)***	38 (0.05)***	38 (0.04)***	38 (0.05)***
C. Depreciation on implements and farm implement buildings	1689 (3)***	8802 (14)***	13733 (20)***	24157 (31)***	10969 (17)***
D. Interest on fixed capital @ 12%	5347 (11)***	6635 (11)***	7336 (11)***	8432 (11)***	6855 (11)***
Amortized establishment costs	16579 (33)***	18952 (30.53)***	20288 (29.55)***	18722 (23.74)***	18894 (29)***
Total Cost (TC)	**142280**	**162441**	**176609**	**191710**	**166697**
Output [Parchment (kg/ha)[a]]	969	1010	1000	1023	1004
Price received/kg	163.23	161.50	163.50	164.48	163.68
Total revenue (TR) (₹/ha)	158129	163128	163500	168334	164335
TR-TC	15848	687	-13109	−23376	-2362
TR-TOC	65897	62772	55536	55486	61776

Source: Field survey.

Note: *, ** and *** figures in parenthesis indicates total cost of cultivation, total operational costs and fixed costs, respectively.

[a] In the case of Arabica parchment and Robusta, cherry coffee was taken into account.

Table 6.2a Net returns per hectare from Robusta variety of coffee in 2015–16 (₹/ha)

Items	*Small (<2 ha)*	*Semi-medium (2–10 ha)*	*Medium (10–25 ha)*	*Large (>25 ha)*	*Overall*
1. Total Operational Cost (TOC)	**83943 (65)***	**91712 (65)***	**93984 (63)***	**105602 (61)***	**92524 (64)***
A. Labour Costs					
1. Hired	14506	23569	41112	53147	28827
2. Family	35612	26533	8364	3712	21828
Sub Total	**50118 (60)****	**50102 (55)****	**49476 (53)****	**56859 (54)****	**50655 (55)****
B. Input Costs					
1.Manures and fertilizers	16889	19909	19551	21250	19569
2. Liming and dolomite	296	441	431	489	425
3. Irrigation	5675	6484	7103	7592	6610
4. Weeding	553	2444	2425	1875	2131
5. Pesticides	1088	1126	1109	1171	1136
Sub Total	**24600 (29)****	**30404 (33)****	**30619 (33)****	**32377 (31)****	**29871 (32)****
C. Maintenance Costs					
2. Transplanting and gap filling	139	72.00	146	166.00	105
4. Fence keeping	234	256	289	350	269
Sub Total	**373 (0)****	**328 (0)****	**435 (0)****	**516 (0)****	**374 (0)****
D. Processing Costs					
1. Processing costs	834	842.00	935.00	904.00	866
2. Costs of materials	314	592.00	583.33	490.00	543
Subtotal	**1148 (1)****	**1434 (2)****	**1518 (2)****	**1394 (1)****	**1409 (2)****
E. Marketing Costs					
1. Transportation charges	634	647	719	1154	710
2. Storage charges	0	0	147	137	43.2

(Continued)

Table 6.2a Continued

Items	*Small (<2 ha)*	*Semi-medium (2–10 ha)*	*Medium (10–25 ha)*	*Large (>25 ha)*	*Overall*
2. Labour charges	171	181	190	185	182
3. Value of bags	2109	2157	2395	2883	2270
Sub Total	**2914**	**2985**	**3451**	**4359**	**3205**
	(3)**	**(3)****	**(4)****	**(4)****	**(3)****
F. Working capital of Robusta growers (₹)					
Total Working Capital (A + B + C + D + E-family labours)	43541	58720	77136	91793	63686
G. Interest on working capital at 11%	**4790**	**6459**	**8485**	**10097**	**7005**
	(6)**	**(7)****	**(9)****	**(10)****	**(8)****
II. Total Fixed Costs	**45939**	**49716**	**56197**	**67510**	**52287**
	(35)*	(35)*	(37)*	(39)*	(36)*
A. Rent on owned land (1/6th value of produce)	29742	31071	30860	30011	30745
	(64.74)***	(62.50)***	(54.91)***	(44.45)***	(58.80)***
B. Land revenue, cesses and taxes	38	38	38	38	38
	(0.08)***	(0.08)***	(0.07)***	(0.06)***	(0.07)***
C. Depreciation on implements and farm implement buildings	3333	4008	9642	20370	6681
	(7.26)***	(8.06)***	(17.16)***	(30.17)***	(12.78)***
D. Interest on fixed capital @ 12%	4915	5318	6012	7224	5594
	(10.70)***	(10.70)***	(10.70)***	(10.70)***	(10.70)***
Amortized establishment costs	7911	9281	9645	9867	9229
	(17.22)***	(18.67)***	(17.16)***	(14.62)***	(17.65)***
Total Cost (TC)	**129881**	**141428**	**150181**	**173111**	**144811**
Output Cherry(kg/ha)	3150	3176	3258	3175	3180
Price received/kg	57	57.04	59	58.33	58.07
Total Revenue (TR) (₹/ha)	179550	181159	192222	185198	184663
TR-TC	49669	39731	42041	12086	39852
TR-TOC	95607	89447	98238	79596	92139

Source: Field survey.
Note: *, ** and *** figures in parenthesis indicates the total cost of cultivation, total operational costs and fixed costs, respectively

Table 6.3a Estimated cost of cultivation and production of coffee on different cost concepts (2015–16)[a]

	Arabica					*Robusta*				
	Small	*Semi-medium*	*Medium*	*Large*	*Overall*	*Small*	*Semi-medium*	*Medium*	*Large*	*Overall*
Cost of Cultivation (₹/ha)										
Cost A1	67466	98621	132440	150711	108712	59612	78506	104945	132164	86640
Cost A2	67466	98621	132440	150711	108712	59612	78506	104945	132164	86640
A2+ Family Labour	110538	128148	142022	155764	132460	95224	105039	113309	135876	108468
B1	72813	105256	139776	159143	115567	64527	83824	110957	139388	92234
B2	99209	132914	167027	186657	142949	94269	114895	141817	169399	122979
C1	115885	134783	149358	164196	139315	100139	110357	119321	143100	114062
C2	142280	162441	176609	191710	166697	129881	141428	150181	173111	144811
Cost of Production (₹/qt)										
A1	6964	9764	13244	14726	10828	1892	2472	3221	4163	2725
A2	6964	9764	13244	14726	10828	1892	2472	3221	4163	2725
A2+ Family Labour	11410	12687	14202	15220	13193	3023	3307	3478	4280	3411
B1	7516	10421	13978	15550	11511	2048	2639	3406	4390	2900
B2	10241	13159	16703	18238	14238	2993	3618	4353	5335	3867
C1	11962	13344	14936	16044	13876	3179	3475	3662	4507	3587
C2	14687	16082	17661	18732	16603	4123	4453	4610	5452	4554
C3	16156	17703	19427	20605	18264	4536	4898	5071	5998	5009

Source: Field survey.

[a] Cost A_1 = Actual expenses in cash + amortized establishment cost and depreciation of fixed assets.
Costs A_2 = Cost A_1 + rent for leased-in land.
Cost B_1 = Cost A_1 + interest on value of owned capital assets (excluding land).
Cost B_2 = Cost B_1 + rental value of owned land and rent paid for leased-in land.
Cost C_1 = Cost B_1 + imputed value of family labour.
Cost C_2 = Cost B_2 + imputed value of family labour.
Cost C_3 = Cost C_2 + 10 per cent of C_2 on account of managerial functions performed by the grower.

Table 6.4a Returns from cultivation of coffee on sample farms (₹/ha)

	Arabica (₹/ha)					*Robusta (₹/ha)*				
Particulars	*Small*	*Semi-medium*	*Medium*	*Large*	*Overall*	*Small*	*Semi-medium*	*Medium*	*Large*	*Overall*
Returns over variable cost	90663	64507	31060	17623	55623	119938	102653	86104	53033	98023
Farm business income	90663	64507	31060	17623	55623	119938	102653	86104	53033	98023
Family labour income	58920	30214	−3527	−18323	21386	85281	66264	49232	15798	61684
Net Income	15848	687	−13109	−23376	−2362	49669	39731	42041	12086	39856
Returns per rupee	1.11	1.00	0.93	0.88	0.99	1.38	1.28	1.27	1.07	1.28

Source: Field survey.

Table 6.5a Gross income vs resources use in coffee production (Y = gross income (₹/ha)

	Arabica (N = 60)			*Robusta (N = 60)*		
Variables	*Coefficient*	*t stat*	*p value*	*Coefficient*	*t stat*	*p value*
Constant	2.77	1.60	0.11	6.72	5.65***	6.08*E*−07
Ln labour use (man days/ha)	0.28	3.09***	0.00	0.18	3.75***	0.00
Ln fertilizers (kg/ha)	0.39	2.67***	0.00	0.15	1.62*	0.10
Ln Irrigation (₹/ha)	0.01	4.02***	0.00	0.21	2.64*	0.01
Ln Pesticides (barrels/ha)	0.23	2.42***	0.01	−0.10	−1.39	0.16
Ln Weedicides (litres/ha)	−0.007	-0.70	0.48	0.00	1.11	0.26
	R^2 = 0.57			R^2 = 0.48		
	F ratio = 14.39***			*F* ratio = 10***		

Source: Field survey.
Note: *, ** and *** represent significance at 10, 5 and 1 per cent, respectively.

Table 7.1a Mode of sales among the surveyed coffee growers in Kodagu during 2016–17 (%)

	Arabica growers					*Robusta growers*				
	Small (n = *8)*	*Semi-medium* (n = *31)*	*Medium* (n = *13)*	*Large* (n = *8)*	*Total* (N = *60)*	*Small* (n = *8)*	*Semi-medium* (n = *34)*	*Medium* (n = *12)*	*Large* (n = *6)*	*Total* (N = *60)*
Local agents at farm gate	37	26	31	12	31	75	52	58	0	52
Coffee depots	63	29	30	0	30	0	0	0	0	0
Curing works	0	32	23	50	23	25	18	8	34	18
Hullers	0	0	0	0	0	0	21	17	0	15
Companies	0	13	8	25	12	0	9	17	33	12
Online sales	0	0	8	0	2	0	0	0	0	0
Direct exports	0	0	0	13	2	0	0	0	33	3

Source: Field survey.

Table 7.2a Variations in monthly farm gate prices of Arabica Parchment in Madikeri over years (₹/50 kg)

Arabica Parchment								
Year	*Standard deviation of prices*	*Coefficient of variation (%)*	*Q1 (Jan-Mar) (₹/50kg)*	*Q2 (Apr-Jun) (₹/50kg)*	*Q3 (Jul-Sep) (₹/50 kg)*	*Q4 (Oct-Dec) (₹/50kg)*	*Highest price (month)*	*Lowest price (month)*
2007	190.64	4.38	4570	4232	4179	4418	July	Jan
2008	215.60	4.38	4752	4810	5205	4935	Jan	Sep
2009	768.81	11.90	5350	6746	7164	6577	Jan	Aug
2010	734.58	10.55	6671	6493	6902	7796	Mar	Dec
2011	348.54	3.46	9734	10377	10072	10061	Jan	May
2012	1055.46	13.12	9354	7931	7994	6904	Nov	Jan
2013	320.91	4.96	6773	6497	6535	6088	Nov	Jan
2014	1688.05	17.22	8133	10558	9687	10841	Jan	Oct
2015	898.01	9.65	10321	9737	9029	8121	Dec	Jan
2016	204.43	2.49	8221	8004	8133	8483	May	Oct

Source: Author's calculations based on secondary data provided by Coffee Board.
Source: Field survey.

Table 7.3a Variations in monthly farm-gate prices of Robusta cherry in Madikeri (₹/50 kg)

Robusta Cherry								
Year	*Standard deviation of prices*	*Coefficient of variation (%)*	*Q1 (Jan-Mar) (₹/50kg)*	*Q2 (Apr-Jun) (₹/50 kg)*	*Q3 (Jul-Sep) (₹/50 kg)*	*Q4 (Oct-Dec) (₹/50 kg)*	*Highest price (month)*	*Lowest price (month)*
2007	135.36	7.45	1683	1722	1922	1943	Nov	Mar
2008	157.06	6.89	2119	2262	2461	2280	Aug	Jan
2009	79.62	4.13	2036	1922	1904	1855	Jan	Dec
2010	183.76	9.83	1812	1658	1907	2103	Nov	Apr
2011	144.67	5.56	2401	2606	2685	2709	Nov	Jan
2012	305.70	10.07	2593	3042	3296	3214	Sep	Feb
2013	106.72	3.49	3102	3064	3143	2917	Jan	Nov
2014	235.94	7.05	3208	3471	3271	3446	Mar	Jul
2015	231.30	7.61	3325	3125	2946	2767	Jan	Oct
2016	278.67	9.18	2817	2850	3004	3471	Dec	Apr

Source: Author's calculation based on secondary data provided by Coffee Board.

Table 7.4a Variations in monthly farm-gate prices of Arabica cherry in Madikeri (₹/50 kg)

Arabica Cherry								
Year	*Standard deviation of prices*	*Coefficient of variation (%)*	*Q1 (Jan-Mar) (₹/50 kg)*	*Q2 (Apr-Jun) (₹/50 kg)*	*Q3 (Jul-Sep) (₹/50 kg)*	*Q4 (Oct-Dec) (₹/50 kg)*	*Highest price (month)*	*Lowest price (month)*
2007	136.52	5.15	2642	2655	2537	2765	Dec	Jul
2008	126.58	4.90	2483	2502	2710	2634	Jul	Apr
2009	419.69	13.45	2642	2655	2537	2765	Dec	Mar
2010	419.69	13.45	2986	2878	3062	3558	Dec	Mar
2011	153.33	3.26	4733	4655	4589	4849	Dec	Aug
2012	425.75	10.24	4656	4087	4156	3727	Jan	Nov
2013	273.94	7.98	3673	3502	3496	3054	Jan	Nov
2014	708.42	15.71	4063	4904	4200	4871	Mar	Jan
2015	529.61	12.50	4796	4541	4112	3496	Jan	Nov
2016	304.32	8.26	3492	3404	3775	4058	Sep	Apr

Source: Author's calculation based on secondary data provided by the Coffee Board.

Table 7.5a Moisture requirement as specified by the Coffee Board (%)

Type of coffee	*Moisture content*
Arabica/Robusta parchment	10 per cent
Arabica cherry	10.50 per cent
Robusta cherry	11 per cent

Source: India Coffee Board.

Table 7.6a Per cent of coffee estates certified under various programs (2016–17)

Size of grower	*Arabica (N = 14)*				*Robusta (N = 15)*			
	Utz (n = 9)	*RFA (n = 4)*	*Nespresso (n = 1)*	*Total (N = 14)*	*Utz (n = 11)*	*4C (n = 2)*	*Organic (n = 2)*	*Total (N = 15)*
Small	0	0	0	0	0	0	0	0
Semi-medium	22.22	0	0	14.29	63.64	50	0	53.34
Medium	55.56	50	100	57.14	27.27	50	50	33.33
Large	22.22	50	0	28.57	9.09	0	50	13.33

Source: Field survey.[a]

[a]Note: All figures refer to share (%) of farmers to total sample size. For instance, 22.22 indicates that 22.22 per cent of 9 Utz certified Arabica growers. *N* refers to total sample size.

Table 7.7a Production vs sales of certified coffee

Certified coffee actually sold as certified (%)[a]		
	Arabica	*Robusta*
Utz	44.44	27.27
RFA	50.00	0.00
Nespresso	100	100
4C	0.00	100
Overall	50.00	46.67

Source: Field survey.

[a] Rest of the certified coffee was sold as ordinary coffee in conventional market.

Table 7.8a Trends in sustainable coffee production and sales at a global level during 2008–12 (MT)

Sustainable coffee Production Vs Sales			
Year	*Production (MT)*	*Sales (MT)*	*Ratio*
2008	1300000	410000	1:03
2009	1800000	440000	1:04
2010	2200000	510000	1:04
2011	2700000	630000	1:04
2012	3300000	840000	1:04
Average	2260000	566000	1:04

Source: Indian Coffee Board, 2017; International Coffee Organization, 2013 and Potts et al., 2014.

Table 7.9a Descriptive statistics on Farm Gate prices and Inter-national prices over the years (2000–17)

	Prices (₹/50 kg)				
	AP (Farm Gate Price)	*AC (Farm Gate Price)*	*RC (Farm Gate Price)*	*Arabica (International Price)*	*Robusta (International Price)*
Average	5671	2744	2000	7531	4030
Minimum	1800	788	444	2991	1257
Maximum	11037	5313	4075	16541	6723
SD	2810	1299	994	3104	1587

Source: Indian Coffee Board, 2017.

Table 7.10a Harvest and post-harvest methods followed by coffee growers (%)

Harvest and post-harvest methods followed by coffee growers (%)

	Harvest and post-harvest operations	**Arabica** Small (n = 8)	Semi-med (n = 31)	Med (n = 13)	Large (n = 8)	Overall (N = 60)	**Robusta** Small (n = 8)	Semi-med (n = 34)	Med (n = 12)	Large (n = 6)	Overall (N = 60)
1	***Type of Harvest***										
	(i) Selective Harvesting	0	9.68	15.38	0	8.33	0	0	0	0	0
	(ii) Strip Harvesting	25	9.68	0	0	8.33	75	85.29	41.67	50	71.67
	(iii) Selective + Stripping	75	80.65	84.62	100	83.33	25	14.71	58.33	50	28.33
	Harvest mats and mixing ripe & unripe berries										
2.	**Use of harvest mats**	25	48	31	100	48	50	88	100	100	87
3.	(i) **Mix ripe & unripe berries**	0	0	0	0	0	100	88	75	0	78
	(ii) No mix ripe & unripe berries	100	100	100	100	100	0	12	25	100	22
	Gleaning activities										
4.	**Undertake Gleaning**	0	39	54	88	43	50	53	50	100	57
	(i) Immediately	0	83	29	100	73	75	64	50	67	62
	(ii) After 1 week	0	0	57	0	15	25	24	50	33	38
	(iii) After 2 weeks	0	17	14	0	12	0	12	0	0	0
5.	**Processing method**										
	(i) Dry method	0	0	0	0	0	100	100	91.67	50	93.33
	(ii) Wet method	100	100	100	100	100	0	0	8.33	50	6.67
6.	**Heaping method**										
	Heaping before drying	88	90	77	75	85	100	88	100	67	90
	1 day heaping	57	64	10	50	51	50	47	83	75	57

(Continued)

Table 7.10a Continued

Harvest and post-harvest methods followed by coffee growers (%)

	2 to 3 days heaping	43	36	80	50	47	0	50	17	25	33
	4 to 5 days heaping	0	0	10	0	2	50	3	0	0	10
	Racking method										
7.	Racking	88	94	100	100	95	100	94	92	100	95
	No racking	12	6	0	0	5	0	6	8	0	5
	Frequency of racking/ day	0	0	0	0	0	0	0	0	0	0
	2 racking/day	0	19	0	38	15	50	74	42	17	58
	3 racking/day	88	74	85	50	75	50	26	50	67	39
	4 racking/day	12	7	15	12	10	0	0	8	17	3
	Drying method										
	(i) Mud/Cow dung yard	37.50	3.23	0	0	6.67	50	5.88	0	0	10
	(ii) Cement/tiled	62.50	96.77	100	100	93.33	50	94.12	100	100	90
	Store Houses										
8.	Store Houses	25	94	100	100	86.67	37	82	92	100	80
	No store houses	75	6	0	0	13.33	63	18	8	0	20

Source: Field survey.

Table 7.11a Quantity (MT) of coffee exported at various quarters

Months of shipment	*2015 Quantity (MT)*	*2016 (Quantity MT)*
1st Quarter	**82902**	**94847**
January	17752	22040
February	27692	31306
March	37458	41501
Per cent to total	27.08	26.42
2nd Quarter	**90215**	**111518**
April	31260	38183
May	29225	36891
June	29730	36444
Per cent to total	29.47	31.07
3rd quarter	**69085**	**75452**
July	25276	24636
August	22974	26591
September	20835	24225
Per cent to total	22.57	21.02
4th Quarter	**63912**	**77128**
October	20812	25956
November	19228	26223
December	23872	24949
Per cent to total	20.88	21.49

Source: Indian Coffee Board, 2017.

Table 7.12a Importing countries and maximum permissible limits (ppb) of OTA

Importing countries	*Green*	*Roasted*	*Instant*
Greece	20	*	*
Hungary	15	10	10
Czech Republic	10	10	10
Italy	8	4	4
Portugal	8	4	4
Spain	8	4	4
Finland	5	5	5
Switzerland	5	5	5
Japan	5	*	*
Germany	*	3	6
Netherlands	*	10	10
European Union	*	5	10

Source: Anon, 2005.

Table 7.13a Shipment of Indian coffee by types and value addition (Quantity in MT)

	Type of coffee				*Value added coffee*			
Year	*Arabica Plantation*	*Arabica Cherry*	*Robusta Parchment*	*Robusta Cherry*	*Instant*	*Roasted*	*Ground*	*Total*
2007–08	30519.20	11595.10	20431.50	86337.20	64857.30	54.10	360.90	214155.30
	(14.25)	(5.41)	(9.54)	(40.32)	(30.29)	(0.03)	(0.17)	(100)
2008–09	38870.90	9999.00	17980.90	89150.10	56547.60	82.20	174.90	212805.60
	(18.27)	(4.70)	(8.45)	(41.89)	(26.27)	(0.04)	(0.08)	(100)
2009–10	19751.20	9890.00	14482.60	79723.20	56188.10	95.50	167.70	180298.30
	(10.95)	(5.49)	(8.03)	(44.22)	(31.16)	(0.05)	(0.09)	(100)
2010–11	38060.10	12520.30	31978.10	120625.30	75401.40	100.30	152.00	278837.50
	(13.65)	(4.49)	(11.47)	(43.26)	(27.04)	(0.04)	(0.05)	(100)
2011–12	40872.40	11738.80	30160.40	158397.50	83383.40	69.70	196.50	324818.70
	(12.58)	(3.61)	(9.29)	(48.76)	(25.67)	(0.02)	(0.06)	(100)
2012–13	40182.30	12375.50	22737.50	125711.30	83033.60	28.30	162.50	302617.00
	(13.28)	(4.09)	(7.51)	(41.54)	(27.44)	(0.01)	(0.05)	(100)
2013–14	43888.29	9730.53	28076.06	134268.48	84972.89	51.97	184.39	301172.60
	(14.57)	(3.23)	(9.32)	(44.58)	(28.21)	(0.02)	(0.06)	(100)
2014–15	44803.35	14392.93	22316.08	111354.60	94369.57	52.79	269.22	287558.54
	(15.58)	(5.01)	(7.76)	(38.72)	(32.82)	(0.02)	(009)	(100)
2015-–16	32776.91	11165.79	35068.80	132484.75	93887.93	160.85	568.84	306113.87
	(10.71)	(3.65)	(11.46)	(43.28)	(30.67)	(0.05)	(0.19)	(100)
2016–17	38493.95	13102.48	28761.01	173119.78	105021.08	62.53	386.70	358947.52
	(10.72)	(3.65)	(8.01)	(48.23)	(29.26)	(0.02)	(0.11)	(100)

Source: Indian Coffee Board, 2017.
Note: Figures in parentheses indicate percentage to total.

Table 7.14a Codex Alimentarius maximum residual limits (MRL) for 21 pesticides

Sl. No	*Pesticides*	*MRL (mg/kg)*
1.	Propiconazole	0.02
2.	Haloxyfop	0.02
3.	Spirodiclofen	0.03
4.	Boscalid	0.05
5.	Chlorpyrifos	0.05
6.	Clothianidin	0.05
7.	Cypermethrin	0.05
8.	Permethrin	0.05
9.	Phorate	0.05
10.	Terbufos	0.05
11.	Tebuconazole	0.1
12.	Aldicarb	0.1
13.	Carbendazim & benomyl	0.1
14.	Disulfoton	0.2
15.	Endosulfan	0.2
16.	Thiamethoxam	0.2
17.	Pyraclostrobin	0.3
18.	Triadimefon	0.5
19.	Triadimenol	0.5
20.	Carbofuran	1
21.	Imidacloprid	1

Source: International Coffee Organization, 2013.

Table 8.1a Market reforms in the coffee sector

Year	*Reforms undertaken*
1989	Collapse of ICA negotiations, resulting in decline in coffee prices
1992	Introduction of ISQ (Internal Sales Quota), wherein Coffee Board allowed the growers to sell 30 per cent of their produce in domestic markets
1994	Introduction of Free Sales Quota (FSQ), wherein the producers could market 50 per cent of their produce either in domestic or export markets
1995	FSQ of small holders increased up to 100 per cent, while large holders increased up to 70 per cent
1996	FSQ of large holders increased up to 100 per cent. Elimination of Coffee Board involvement in coffee marketing and monopoly of the Coffee Board was broken. Entry of private traders into marketing of coffee. In 1996, the Coffee Board was disintegrated and monopoly of the Board was broken down. This was a perfect example of institutional failure in the coffee system. With the 'hard institutional failure' (failure pertaining to regulations and the legal system), the role of institution (Coffee Board) after liberalization was restricted to research, development and promotional activities.

Source: Indian Coffee Board.

Table 8.2a Arabica varieties of coffee cultivated among the growers

Arabica variety	*Marginal* (n = *8)*	*Small* (n = *31)*	*Medium* (n = *13)*	*Large* (n = *8)*	*Overall* (n = *60)*
S.795	75.00	83.87	84.62	100.00	85.00
Sln.6	75.00	54.84	38.46	12.50	48.33
Sln. 9	50.00	19.35	38.46	37.50	30.00
Catuai	37.50	22.58	15.38	0.00	20.00
Chandragiri	12.50	22.58	15.38	25.00	20.00
Cauvery	0.00	16.13	15.38	0.00	11.67
Kents	0.00	6.45	7.69	0.00	5.00
Sln.5	0.00	0.00	0.00	12.50	1.67

Source: Field survey.

Table 8.3a Extension activities conducted by the Coffee Board to control Coffee Berry Borer

Extension activities by Coffee Board	*1998–99*	*1999–2000*	*2000–01*	*2001–02*	*Total as on 31/08/2001*
Visits, tours and gatherings					
Contact visits	9811	3442	11671	3546	28470
Visit to IPM Plots	50	52	52	52	206
Group gatherings at IPM plots	0	1	17	1	19
Study tours	0	0	43	0	43
Workshops and seminars					
Technical workshops	3	6	18	8	35
Farmers' Group workshops	0	0	240	153	393
District level regional review workshops	0	1	8	2	11
State level Apex workshops	0	0	1	1	2
Seminars on CBB	3	1	12	4	20
Media campaigns					
Newspaper	15	5	13	5	38
Radio announcements	19	1	4	4	28
Radio talks	4	3	5	1	13
Others					
Issue of advisory letters on CBB	1925	331	1729	466	4451
Contact/assessment camps	13	3	11	1	28

Source: Bentley et al., 2002.

Table 8.4a Mean number of Coffee Berry Borers trapped/day in Waynad (Jan–Feb 2000)

Mean number of Coffee Berry Borers caught/trap		
Type of trap	*Harvested field/day*	*Non-Harvested field/day*
Multi-funnel trap	668	72
Plastic-can trap	494	1921
Single-funnel trap with reservoir	-	11

Source: Bentley et al., 2002.

Table 8.5a Major certification schemes for coffee

Certification schemes	*Objective*	*Standard setting body*	*Price premiums*
Organic coffee	Develops a verified sustainable coffee system in harmony with nature, promotes bio diversity and natural soil health	IFOAM (International Federation of Organic Agriculture Movements)	USD $0.04--0.30 per pound
Fair-trade coffee	Improving livelihoods of small producers, improved market access, fair prices, direct trade and environmental stewardship	Fair-trade Labelling Organizations International	Minimum price of USD$ 1.25 per pound, addition of UD$ 0.30 per pound of organic coffee besides USD$ 0.20 per pound of social premium.
Rain forest Alliance (RFA) coffee	Bio-diversity conservation, sustainable farm management and community development	Sustainable Agriculture Network	No guaranteed minimum price
Utz certified coffee	Maintains sustainable supply chain and traceability issues, provides a holistic approach for coffee production	Utz Kapeh	USD$ 0.07 per pound
4C (Common Codes for the Coffee Community) **	Sets up baseline standards for sustainability in coffee. Enhances good practices for better livelihoods. Business to business without labelling	4C Association	No individual negotiation.

Source: Adapted from Kline (2010) and Ponte (2004).

** Eco-labelling of '4C Logo' is not seen in the marketing of coffee. The 4C concept does not believe in building up a 'Brand name' for communicating to consumers but believes in improving the sustainable practices in coffee.

Index

Page numbers in **bold** and *italics* refer to tables and figures.

For Product Safety Concerns and Information please contact our EU
representative GPSR@taylorandfrancis.com
Taylor & Francis Verlag GmbH, Kaufingerstraße 24, 80331 München, Germany

www.ingramcontent.com/pod-product-compliance
Ingram Content Group UK Ltd.
Pitfield, Milton Keynes, MK11 3LW, UK
UKHW020149200926
475702UK00016B/2623

* 9 7 8 1 0 3 2 8 9 2 0 5 4 *